TECHNOLOGY ALONE, IS NOT TRANSFORMATION

THE HR TECHNOLOGIST'S GUIDE
TO TRUE ORGANIZATIONAL CHANGE

LEE CAGE JR.

Book Design by Molly Seabrook
Edited by Traci Cuthbertson

ISBN 979-8-9928986-1-3

Published by Lee Cage, Jr.

www.leecagejr.com

TABLE OF CONTENTS

INTRODUCTION

The Technology Trap

As I powered down my laptop after yet another video call with a frustrated Chief People Officer, I couldn't help but reflect on the pattern I'd observed across dozens of organizations. The CPO had just spent millions on a state-of-the-art HR management system, yet employee engagement was down, managers were complaining about increased administrative burden, and the promised efficiencies hadn't materialized. Sound familiar?

"We've implemented the best technology money can buy," she had lamented. "Why aren't we seeing transformation?"

In my twenty years at the intersection of HR and technology, I've witnessed this scene play out repeatedly—organizations investing heavily in digital solutions only to find themselves with expensive new problems rather than the revolutionary changes they expected. The hard truth that many leaders struggle to accept is simple yet profound: Technology alone is not transformation.

Let me be clear: I'm a technologist at heart. I've built HR platforms, advised Fortune 500 companies on technology implementation, and evangelized digital approaches to people management throughout my career. Technology remains a powerful enabler of change—but mistaking the tool for the journey is where many organizations go wrong.

This book is for HR leaders who find themselves seduced by sleek interfaces and AI-powered promises yet frustrated by the gap between technological implementation and true organizational transformation.

It's for the professionals who intuitively understand that lasting change requires more than a software upgrade but struggle to articulate and implement that understanding in a world obsessed with digital solutions.

Throughout these chapters, we'll explore what true transformation looks like beyond the digital veneer. We'll examine how successful organizations integrate technology within a broader change ecosystem—one that encompasses human psychology, organizational culture, leadership models, and systemic thinking. Most importantly, we'll reframe the role of HR not as technology administrators but as architects of transformative experience.

As an HR leader armed with a technologist's mindset, you stand at a unique intersection—understanding both human potential and technological capability. This perspective positions you perfectly to lead genuine transformation efforts that leverage technology as one component of a more complex, more human-centered approach to organizational evolution.

The pages that follow contain no silver bullets or quick fixes. What you will find is a framework for thinking beyond the technology trap and embracing a more nuanced, more effective path to meaningful change. If you're ready to look beyond the shiny digital objects and build something truly transformative, then let's begin.

1

The Transformation Mirage: Why Technology Isn't Enough

CHAPTER 1

The Transformation Mirage: Why Technology Isn't Enough

The email landed in my inbox with predictable enthusiasm: "Exciting news! We're implementing a new performance management system that will transform how we develop talent!"

Six months later, I was sitting across from the same HR director who had sent that email, now slumped in her chair, explaining how managers were finding workarounds to avoid using the system, employees were complaining about the additional clicks, and the promised insights remained elusive. Technology had been deployed, but transformation was nowhere to be found.

This scenario illustrates what I call the "Transformation Mirage", the persistent and costly illusion that implementing new technology is synonymous with transforming an organization. Let's examine why this mirage persists and why technology alone consistently falls short of delivering true transformation.

The Digitization-Transformation Confusion

First, we must distinguish between digitization and transformation. Digitization is converting analog processes to digital ones, moving paper forms online or automating manual workflows. This is valuable but ultimately represents a change in medium, not substance.

Transformation, by contrast, involves fundamentally rethinking how work happens, often challenging basic assumptions about processes, relationships, and value creation. When we merely digitize without transforming the underlying models, we simply make our existing limitations more efficient.

Consider the classic example of early e-learning platforms that essentially took classroom lectures and put them online. The result? High dropout rates and minimal engagement. The technology changed, but the learning paradigm—passive consumption of information—remained unchanged. True transformation came later when learning platforms incorporated social learning, microlearning, adaptive pathways, and experiential components that fundamentally reimagined how learning happens.

The Implementation Fallacy

Organizations frequently fall victim to what I call the "Implementation Fallacy", the belief that once technology is successfully installed and functioning according to specifications, transformation will naturally follow. This perspective dangerously reduces transformation into a technical challenge rather than the complex social, psychological, and organizational change it represents.

Implementation focuses on getting systems up and running. Transformation focuses on creating new capabilities, behaviors, and outcomes. One is an event; the other is a journey.

A large manufacturing company I worked with spent $12 million implementing a workforce management system to optimize scheduling and productivity. The implementation was technically flawless, as the system worked exactly as designed. But when I visited six months later, productivity had barely budged. Why? Managers continued making decisions the same way they always had, simply inputting their conclusions into the new system rather than using its analytical capabilities to inform better decisions. Technology was present, but the transformation in decision-making was missing.

The Ecosystem Blindness

Technology never exists in isolation—it operates within an ecosystem of people, processes, skills, incentives, and cultural norms. When we focus solely on the technology component, we develop ecosystem blindness, failing to see how these other elements must also evolve for transformation to occur.

A financial services firm implemented an impressive internal talent marketplace platform to increase internal mobility and development. Technology worked beautifully, but adoption languished. Further investigation revealed that:

- Manager performance was still evaluated on team retention metrics, creating a disincentive to support team members moving to new roles

- Employees feared that expressing interest in other positions would signal disloyalty to their current leaders

- The organizational culture still implicitly values specialization over diverse experiences

- No process existed for transitioning work when someone took a new internal role

Technology couldn't overcome these ecosystem factors alone. True transformation required aligning incentives, cultural values, processes, and mindsets with the new technical capabilities.

The Why-How Disconnect

When organizations become enamored with technology, they often start with "how" (how will we implement this tool?) before clarifying "why" (why are we doing this and what outcomes do we seek?). This disconnects leads to technically successful implementations that fail to deliver meaningful change.

I've observed countless HR teams who can explain in detail how their new recruiting system works, the workflows, data capture points, and automation features—but struggle to articulate how these capabilities connect to the fundamental talent acquisition challenges their organization faces. Without this clarity, technology becomes a solution in search of a problem rather than a targeted tool for transformation.

The Human Adaptation Gap

Perhaps most fundamentally, technology-centric approaches to transformation often underestimate the human adaptation required. New technologies demand new behaviors, skills, relationships, and ways of thinking. This adaptation doesn't happen automatically, it requires intention, support, and time.

When a global professional services firm implemented a sophisticated collaboration platform, they reasonably expected it would break down silos between practice areas. But even with perfect functionality and thorough training, they discovered that cross-practice collaboration involves much more than technology:

- Consultants needed psychological safety to share half-formed ideas
- Team leaders needed to develop skills in facilitating virtual collaboration
- Practice incentive structures needed realignment to reward cross-boundary work
- Professionals needed time to build trust with colleagues they rarely saw in person

Technology provided a venue for collaboration, but the human adaptation required for true transformation extended far beyond learning to use the platform.

From Mirage to Reality

If technology alone doesn't bring transformation, then what does?

True transformation emerges from the intentional orchestration of multiple dimensions:

1. Purpose and strategy: Clear articulation of why change is needed and how it connects to organizational purpose

2. Human experience: Designing for the psychological and social realities of those affected

3. Process reimagination: Fundamentally rethinking how work gets done, not just digitizing existing processes

4. Cultural alignment: Ensuring shared values and norms support new ways of working

5. Leadership models: Evolving how direction, alignment, and commitment are created

6. Skills and capabilities: Building new competencies required for success

7. Structural considerations: Adjusting organizational design to enable new work patterns

8. Technology enablement: Deploying tools that support and enhance the above elements

This multidimensional view doesn't diminish technology's importance—it simply positions it as one essential element in a more complex transformation ecosystem.

As HR leaders with a technologist's mindset, we must resist the allure of the transformation mirage. We must champion a more sophisticated understanding of change that acknowledges technology's power while recognizing its limitations. Only then can we move beyond digital disappointment to genuine organizational transformation.

In the chapters that follow, we'll explore each dimension of true transformation in detail, providing frameworks and practical approaches for orchestrating change that delivers on the promise that technology alone cannot fulfill.

2

The Human Element: People as the Core of Transformation

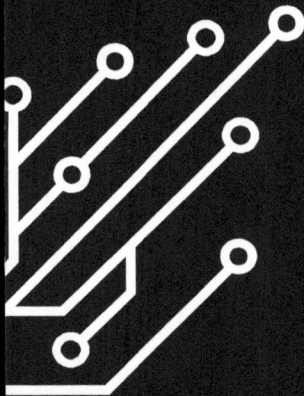

16

CHAPTER 2

The Human Element:
People as the Core of Transformation

"We've built an incredible self-service HR platform with all the bells and whistles," the HR technology director told me proudly. "But we're still seeing 70% of employees calling the service center instead of using it."

This scenario illustrates a fundamental truth that technology-first transformation efforts often overlook: at its heart, organizational transformation is a deep human endeavor. Technology may enable new possibilities, but people ultimately determine whether transformation succeeds or fails through their adoption, resistance, creativity, and commitment.

In this chapter, we'll explore why the human element must be at the core of any transformation effort and how HR leaders can design change initiatives that work with rather than against human psychology.

The Psychology of Change

To understand why technology alone doesn't transform organizations, we must first understand how humans experience and respond to change. Despite our capacity for adaptation, humans are inherently conservation-oriented beings. We develop mental models, routines, and expectations that create efficiency and psychological safety in our work lives.

Technology-driven change initiatives often trigger threat responses because they disrupt these established patterns. When faced with new systems or ways of working, people typically experience:

1. Loss aversion: The pain of losing familiar tools and processes often outweighs the perceived benefit of new ones

2. Identity challenges: Changes in how work happens can threaten professional identity ("I'm the Excel expert" becomes less valuable when analytics are automated)

3. Competence anxiety: Fear of appearing incompetent when learning new systems

4. Ambiguity discomfort: Uncertainty about how new ways of working will affect status, relationships, and performance

5. Value misalignment: Concern that technology-driven changes prioritize efficiency over human values

These psychological responses aren't irrational resistance to be over-come—they're natural human reactions that transformation efforts must address to succeed.

A pharmaceutical company I advised implemented a cutting-edge talent analytics platform but found that hiring managers continued relying on "gut feel" despite having access to predictive data. The transformation team had focused entirely on data quality and user interface while neglecting the psychological shift required—helping managers develop comfort with data-informed (rather than purely intuitive) decision-making and addressing their fears about diminished authority.

From Users to Humans

Technology-centric transformation typically views people as "users" to be trained rather than complex humans navigating change. This perspective leads to an overemphasis on feature adoption and under-investment in the deeper human journey.

Consider two contrasting approaches to implementing a new performance management system:

The Technology-Centric Approach:

- Focus: System functionality and adoption metrics
- Success defined by: Completion rates and technical accuracy
- Change management: Training on "how to use the system"
- Primary question: "Are people using the features correctly?"

The Human-Centric Approach:

- Focus: Evolving the feedback culture and improving development conversations
- Success defined by: Quality of conversations and perceived developmental value
- Change management: Building feedback skills and psychological safety
- Primary question: "Are people having better developmental conversations?"

The second approach recognizes that technology is merely a means to a human end, better conversations and development, rather than an end itself.

Designing for Human Adaptation

True transformation requires designing not just for technological implementation but for human adaptation. This means creating conditions where people can successfully navigate the psychological, social, and practical challenges of change.

When a global retailer transformed its approach to scheduling through a new workforce management system, the most successful stores weren't those with the most technically proficient managers. They were stores where leaders:

1. Created psychological safety by normalizing struggles with the new system

2. Provided social proof by highlighting early adopters and their successes

3. Reduced status threats by having associates help managers learn the system

4. Connected to purpose by emphasizing how better scheduling improved customer service and work-life balance

5. Built new habits through structured practice and peer coaching

These leaders intuitively understood that human adaptation—not technical implementation—was the critical success factor.

The Experience Layer

Transformative technology requires what I call an "experience layer", the thoughtfully designed human experiences that surround and give meaning to technical capabilities. This layer includes:

- Narrative: The story that connects change to purpose and meaning

- Interactions: How people engage with each other around technology

- Environment: Physical and cultural contexts that support new ways of working

- Support structures: Resources for learning and adaptation

- Recognition: How progress and success are acknowledged

A large insurance company implementing a new digital workspace platform initially focused exclusively on technical features. Adoption stalled until they invested in the experience layer—creating digital comfort zones where teams could experiment together, developing peer guides who provided contextual support, and establishing rituals for celebrating digital wins and learning from failures.

The Relationship Revolution

Perhaps the most overlooked aspect of transformation is how technology changes human relationships, between employees and managers, between team members, and between the organization and its people.

When a university moved its performance management process online, they unintentionally disrupted the relationship dynamics between faculty and department chairs. What had been relationship-driven development conversations became form-filling exercises. True transformation occurred only when they redesigned the process to use technology for documentation while preserving and enhancing human conversation at its core.

This example highlights a critical insight: technology should serve relationships, not replace them. The most successful transformations use technology to automate transactional elements while creating space for more meaningful human connections.

From Recipients to Co-Creators

Traditional change approaches the position of people as recipients of transformation designed by others. But humans engage more deeply when they participate in creating change rather than merely receiving it.

A manufacturing organization implementing connected factory technology initially faced resistance until they shifted to a co-creation model. They established innovative teams where frontline workers collaborated with engineers to design how the technology would be integrated into daily work. The result wasn't just higher adoption, but better solutions informed by practical expertise.

This co-creation approach recognizes that transformation isn't something done to people, but something done with them. It leverages the distributed intelligence of the organization while building the ownership essential for sustained change.

The Human Transformation Scorecard

To counterbalance technology-centric metrics, HR leaders need what I call a "Human Transformation Scorecard" that tracks the people's dimensions of change:

1. Psychological measures: Safety, confidence, optimism about the future

2. Behavioral indicators: New practices, collaborative patterns, initiative

3. Relationship quality: Trust, information sharing, conflict resolution

4. Capability development: New skills, adaptability, learning velocity

5. Experience metrics: Effort, friction, moments of delight or frustration

These measures provide a more complete picture of transformation progress than technical adoption metrics alone. They help leaders identify and address the human factors that ultimately determine whether technology delivers its promised value.

Leading with Empathy

Perhaps most importantly, true transformation requires empathetic leadership that acknowledges and addresses the human complexity of change.

This means:

1. Recognizing that resistance often stems from legitimate concerns rather than obstinance
2. Creating psychological safety for people to express difficulties and ask for help
3. Personalizing support based on different adaptation needs and paces
4. Modeling vulnerability by sharing leaders' own challenges with change
5. Celebrating progress while acknowledging the real costs of transition

Leaders who demonstrate this empathetic approach create conditions where people can move through the discomfort of change to embrace new possibilities.

The Human-Technology Partnership

Ultimately, true transformation occurs when technology enhances human capabilities instead of replacing them, creating a partnership that leverages the unique strengths of both.

Consider how the most successful AI implementations in HR don't simply automate human judgment but enhance it. It does so by providing insights that form better human decisions, automating routine tasks to create space for more meaningful work, and connecting people in ways that enhance rather than diminish human relationships.

This partnership perspective moves us beyond the false dichotomy of "high-tech versus high-touch" to a more sophisticated understanding of how technology and humanity can evolve together.

As HR leaders seeking true transformation, our challenge is to become as sophisticated about human adaptation as we are about technological implementation. When we place people at the core of our transformation efforts, designing for their psychology, experiences, relationships, and participation, only then can we create conditions where technology can fulfill its transformative potential.

In the next chapter, we'll explore how organizational culture functions as the operating system for transformation, either enabling or constraining the change technology makes possible.

3

Culture: The Operating System of Transformation

26

CHAPTER 3

Culture: The Operating System of Transformation

In my consulting work, I often ask executives this question: "If I gave you an identical technology stack as your most innovative competitor, would you achieve the same results?"

The answer is invariably no. Even with identical technological capabilities, organizations achieve dramatically different outcomes based on their cultures. The shared beliefs, values, and behaviors determine how work actually happens.

This reality points to a critical insight: culture functions as the operating system for transformation. Just as computer hardware requires an operating system to translate its capabilities into useful functions, technology requires a compatible cultural operating system to deliver transformative value.

In this chapter, we'll explore how culture enables or constrains technological change and how HR leaders can actively shape cultural operating systems that power true transformation.

Culture as Software

Think of technology as organizational hardware and culture as its software. Hardware provides capability, but software determines how that capability is expressed and what it produces. You can upgrade hardware (implement new technology), but if you're running outdated or incompatible software (culture), the system's performance will be severely limited.

I witnessed this dynamic at a financial services firm that implemented a state-of-the-art internal collaboration platform. The technology worked flawlessly, but usage was minimal for non-essential communication. Research showed that the platform's collaborative features were fundamentally incompatible with the cultural operating system, which was marked by information hoarding, competitive internal dynamics, and a dread of documented blunders.

No amount of training or technical refinement could overcome this cultural incompatibility. True transformation required upgrading both the technological hardware and the cultural software simultaneously.

The Cultural Compatibility Assessment

Before implementing new technology, organizations should conduct what I call a "Cultural Compatibility Assessment". This examines alignment between the cultural assumptions embedded in technology and the existing organizational culture.

Consider these examples of potential misalignment:

- Agile project management tools assume transparent sharing of progress and problems but may conflict with cultures where admitting challenges is seen as weakness.

- Innovation platforms assume people will voluntarily contribute ideas but may struggle in cultures where speaking up has historically been risky.

- Self-service HR systems assume employee autonomy but may clash with authoritative cultures where managers expect to control access to information.

- Social recognition tools assume public acknowledgment is motivating, but may create discomfort in cultures that value modesty and private recognition

These misalignments don't mean the technology is inappropriate—but they do signal that cultural evolution must accompany technological implementation for transformation to succeed.

The Cultural Levers of Transformation

When we recognize culture as the operating system for transformation, we can identify and activate specific cultural levers that enable technology to deliver its full value. These levers include:

1. Values in Action

Values statements alone don't determine culture, what matters is how values manifest in daily decisions, especially when under pressure. Organizations that successfully utilize technology for transformation ensure their proclaimed values align with technology-enabled behaviors.

A healthcare system implementing telemedicine faced resistance until leadership explicitly connected the initiative to their core value of patient-centered care, demonstrating how virtual care expanded access for vulnerable populations. This value alignment transformed the perception of technology from an efficiency-driven imposition to an expression of organizational purpose.

2. Leadership Behaviors

Leaders shape culture through what they model, what they measure, and what they celebrate. When leaders' behaviors contradict the collaborative, transparent, or innovative possibilities that new technology enables, transformation stalls.

A professional services firm struggled with adoption of their knowledge-sharing platform until the managing partner began regularly posting questions, acknowledging others' contributions, and openly discussing how shared insights had influenced his thinking. This modeling created psychological safety for others to engage, dramatically increasing platform value.

3. Decision Rights and Autonomy

How decisions are made, who has authority, how input is gathered and how quickly decisions happen, is a critical cultural dimension that affects technological transformation.

A manufacturing company implemented advanced analytics capabilities but saw minimal impact on operations because the culture required decisions to flow through multiple hierarchical layers before action could be taken. Technology produced insights faster than the culture could respond to them. Transformation accelerated only when decision rights were pushed closer to the front line, allowing faster response to analytical insights.

4. Learning Orientation

Organizations with strong learning cultures, characterized by psychological safety, constructive feedback, and viewing failure as learning, leverage technology more effectively for transformation than those with performance-oriented cultures focused on error avoidance.

A technology company implementing a new customer relationship management system initially struggled with adoption until they shifted from a compliance-focused implementation ("You must use the system correctly") to a learning-focused approach ("We're learning together how to better serve customers"). This cultural shift encouraged experimentation and peer learning that accelerated both adoption and innovation.

5. Relationship Networks

Informal networks, who talks to whom, who influences whom, who trusts whom, often determine how effectively transformation spreads through an organization.

A global manufacturer successfully accelerated their digital transformation by mapping relationship networks and deliberately engaging informal influencers as change ambassadors. These respected colleagues provided social proof that reduced resistance more effectively than formal authority could have.

6. Stories and Narrative

The stories an organization tells itself, about its history, its heroes, its future shape how people interpret and respond to change.

A financial institution struggled with their digital transformation until they began collecting and sharing stories of how new technologies were helping employees better serve customers. These narratives shifted the cultural interpretation of technology from "efficiency tool that threatens jobs" to "enabler of more meaningful customer relationships."

Cultural Evolution by Design

Recognizing culture's role in transformation doesn't mean waiting for culture to change organically before implementing technology. Instead, it means deliberately designing cultural evolution alongside technological implementation.

A retail organization I worked with took this integrated approach when implementing AI-powered workforce scheduling. They recognized that technology conflicted with their existing culture in several ways:

- Cultural norm: Managers controlled schedules as a sign of authority
- Technology assumption: Algorithms optimize schedules based on multiple variables

Rather than simply training managers on the system, they designed deliberate cultural interventions:

1. Reframed the manager's role from "schedule creator" to "schedule optimizer and coach"

2. Created new rituals where managers and teams reviewed algorithm suggestions together

3. Established recognition for managers whose teams achieved both efficiency and satisfaction

4. Updated leadership development to emphasize data-informed decision-making

5. Modified performance metrics to balance algorithmic efficiency with team wellbeing

This deliberate cultural evolution allowed technology to deliver its intended value while preserving crucial human judgment and relationship elements.

The Subculture Reality

Large organizations rarely have monolithic cultures. Instead, they comprise of multiple subcultures with different characteristics. This reality explains why the same technology often produces dramatically different results across departments or locations.

A global pharmaceutical company implementing a digital learning platform saw adoption rates ranging from 15% to 85% across divisions. Analysis revealed that high-adoption areas shared cultural characteristics, comfort with experimentation, peer learning norms, and leaders who modeled continuous development, regardless of function or geography.

This insight suggests that transformation strategies should be tailored to subculture realities rather than assuming a one-size-fits-all approach. It also points to the value of identifying and learning from "bright spots", pockets where culture and technology have successfully integrated to produce transformation.

Cultural Debt and Technical Debt

In technological development, "technical debt" accumulates when short-term expediency leads to suboptimal code that must eventually be restructured. Similarly, organizations accumulate "cultural debt" when they implement new technologies without addressing underlying cultural incompatibilities.

A telecommunication company rapidly deployed customer self-service technologies without addressing their internally focused, engineering-driven culture. The result was technically sophisticated systems that customers found frustrating because they reflected internal processes rather than customer needs. Years later, they invested heavily in customer experience transformation, essentially paying down the cultural debt they had accumulated.

This pattern suggests that addressing cultural factors alongside technological implementation isn't just good to change management, it's financial prudence that prevents costly remediation later.

Measuring Cultural Evolution

Traditional transformation metrics focus on technology adoption and technical outcomes. To track cultural evolution, HR leaders need different measures:

- Psychological safety indicators: Willingness to admit mistakes, ask questions, and challenge status quo

- Collaboration patterns: Cross-functional interactions, knowledge sharing, joint problem-solving

- Decision velocity: Time from insight to action across different levels

- Learning behaviors: Experimentation frequency, feedback quality, adaptation speed

- Purpose alignment: Connection between daily work and organizational mission

These cultural metrics provide early indicators of whether the organization is developing the operating system needed for technology to deliver transformative value.

Cultural Catalysts in HR

As administrators of organizational culture, HR leaders are uniquely positioned to serve as catalysts for the cultural evolution required for true transformation. This role involves:

1. Examining cultural assumptions that may help or hinder technological change
2. Designing talent practices that reinforce desired cultural characteristics
3. Advising leaders on cultural implications of technical decisions
4. Creating spaces for experimentation and learning
5. Curating and sharing stories that illustrate cultural evolution in action

When HR fulfills this catalytic role, it helps bridge the gap between technological possibility and cultural reality that often derails transformation efforts.

The Cultural Transformation Canvas

To integrate cultural considerations into transformation planning, I recommend using what I call the "Cultural Transformation Canvas", a tool that prompts leaders to address key questions:

1. Current Culture: What cultural characteristics will help or hinder our technological change?
2. Technology Assumptions: What beliefs and behaviors do our technology assume or require?

3. Gap Analysis: Where do we see significant misalignment between current culture and technology requirements?

4. Cultural Evolution Priorities: Which cultural elements must shift first to enable transformation?

5. Leverage Points: What existing cultural strengths can we build upon?

6. Intervention Design: What specific actions will catalyze the cultural evolution needed?

7. Leadership Alignment: How will leaders model and reinforce desired cultural characteristics?

8. Measurement Approach: How will we track cultural evolution alongside technical implementation?

Organizations that thoughtfully complete this canvas develop more integrated transformation strategies that address both technological and cultural dimensions of change.

Culture as Competitive Advantage

Perhaps most importantly, culture represents the most sustainable competitive advantage in a world where technology is increasingly standardized. Your competitors can purchase the same systems, but they cannot easily replicate a culture that maximizes those systems' value.

A regional bank I advised differentiated itself not through unique digital banking features (which larger competitors could easily match) but through a distinctive culture of tech-enabled personal service. Their competitive advantage came not from what their technology could do but from how their people used that technology to create memorable customer experiences.

This reality suggests that HR leaders should focus not just on implementing the latest technologies but on cultivating cultures that uniquely support those technologies to deliver value that competitors cannot easily copy.

As we move into the next chapter on leadership, remember this essential truth: culture is not a soft factor that follows technological change rather it's the operating system that determines whether that change delivers transformative value. When we invest as thoughtfully in cultural evolution as we do in technological implementation, we create the conditions for true transformation to flourish.

4

The Leadership Paradigm: from Command to Catalyst

CHAPTER 4

The Leadership Paradigm: From Command to Catalyst

The CEO's frustration was palpable. "We've invested millions in digital transformation," she said, "but our middle managers just won't get on board. They're the bottleneck to everything we're trying to do."

This complaint, variations of which I've heard countless times, reflects a fundamental misunderstanding of leadership's role in transformation. The issue isn't the resistant middle managers; it's a leadership model misaligned with the realities of technology-enabled change.

In this chapter, we'll explore how traditional command-and-control leadership impedes transformation and how a shift to catalytic leadership creates the conditions for technology to deliver its transformative potential.

The Leadership-Technology Mismatch

Most organizations implement 21st-century technology while maintaining 20th-century leadership models. This mismatch creates friction that slows down transformation efforts.

Traditional leadership models were designed for industrial-era organizations characterized by:

- Predictable environments and stable competitive landscapes
- Sequential, standardized work processes
- Information scarcity and centralized knowledge
- Clear hierarchical decision rights
- Relatively homogeneous workforces

Modern technologies, from collaboration platforms to AI-powered analytics, create possibilities that these leadership models aren't designed to leverage:

- Rapid experimentation and adaptation
- Networked, cross-functional collaboration
- Distributed intelligence and collective innovation
- Empowered decision-making at all levels
- Diverse perspective integration for complex challenges

A global consumer products company exemplified this mismatch when implementing an innovative platform designed to source ideas from employees worldwide. The technology worked perfectly, but participation languished. Investigation revealed that while technology enabled bottom-up innovation, the leadership model remained firmly top-down. Leaders are expected to control idea flow, evaluate suggestions privately, and maintain decision authority all while undermining the platform's core purpose of distributed innovation.

The Three Leadership Paradigms

To understand the leadership shift required for true transformation, let's examine three paradigms that exist in organizations today:

1. Command Leadership

This traditional paradigm views leaders as authoritative decision-makers who:

- Provide clear direction and control
- Allocate resources based on established priorities
- Ensure compliance with established processes
- Motivate through incentives and consequences
- Maintain boundaries between functions and levels

While command leadership can be effective in stable, predictable environments, it significantly constrains the value technology can deliver in dynamic contexts.

2. Consultative Leadership

This evolved paradigm views leaders as expert guides who:

- Solicit input before making decisions
- Delegate within controlled parameters
- Encourage limited experimentation within boundaries
- Coach teams toward predetermined outcomes
- Create connections across select boundaries

While consultative leadership is more participative than command leadership, this paradigm still positions leaders as primary decision-makers and controllers of change that results in limiting the distributed innovation and adaptation that modern technology enables.

3. Catalytic Leadership

This emergent paradigm views leaders as enablers who:

- Create conditions for others to drive change
- Allocate resources based on emerging opportunities
- Design environments for experimentation and learning
- Inspire through purpose and meaning
- Remove barriers to cross-boundary collaboration

This catalytic paradigm aligns with technology's potential to enable distributed intelligence, rapid adaptation, and collaborative innovation.

A healthcare organization I advised demonstrated this evolution when implementing telehealth technology. Initially, leaders took a command approach, mandating adoption targets and monitoring compliance. Results were disappointing: high technical adaptation but low-quality telehealth experiences.

When they shifted to catalytic leadership, creating practitioner design teams, establishing learning communities, and celebrating patient experience innovations, only then the same technology delivered dramatically better outcomes. The difference wasn't technology but how leadership enabled its use.

The Catalytic Leadership Model

Catalytic leadership isn't about abandoning responsibility but about exercising leadership differently to create conditions where technology-enabled transformation can flourish. This approach encompasses five key practices:

1. Frame Meaningful Challenges

Rather than defining specific solutions, catalytic leaders frame meaningful challenges that:

- Connect to purpose and values
- Have sufficient clarity to provide direction
- Maintain enough openness for diverse approaches
- Create urgency without specifying methods

A manufacturing leader exemplified this practice when implementing proprietary sensors. Rather than mandating specific uses, she challenged teams to "reduce unplanned downtime by 50% using any combination of new sensors and existing knowledge." This challenge orientation unleashed creative applications beyond what central planners could have designed.

2. Create Psychological Safety

Transformation requires experimentation, which inevitably involves setbacks. Catalytic leaders deliberately create psychological safety that enables learning by:

- Normalizing challenges as part of transformation
- Sharing their own mistakes and lessons
- Focusing inquiry on learning rather than blame
- Protecting teams during vulnerable transitions
- Celebrating thoughtful experiments regardless of outcome

A financial services executive demonstrated this when his digital banking platform encountered early issues. Rather than seeking culprits, he convened a "learning circle" where he first shared his own concerns about whether he had provided sufficient resources. This psychological safety enabled the team to identify and address root causes without fear.

3. Balance Structure and Emergence

While traditional leadership emphasizes control through rigid structures, catalytic leadership creates what I call "enabling constraints", frameworks that provide sufficient guidance while allowing for emergence and adaptation.

A retail organization implementing AI-powered inventory management effectively balanced structure and emergence by:

- Establishing clear guardrails for algorithm use
- Creating regular human oversight checkpoints
- Encouraging store-level experimentation within parameters
- Developing mechanisms to scale successful variations
- Maintaining space for human judgment in customer-facing decisions

This balanced approach enabled technology to continuously improve through distributed learning while maintaining essential consistency.

4. Foster Boundary-Spanning Connections

True transformation often requires integrating insights and capabilities across traditional organizational boundaries. Catalytic leaders deliberately foster these connections through:

- Cross-functional transformation teams
- Shared metrics that require collaboration
- Rotation programs that build perspective diversity
- Physical and virtual spaces for unplanned interactions
- Recognition that highlights boundary-spanning work

A pharmaceutical company implementing a digital clinical trial platform created "integration teams" comprising clinical, data science, patient advocacy, and regulatory professionals. These boundary-spanning teams identified integration points those central planners had missed and developed protocols that balanced innovation with compliance requirements, something no single function could have achieved alone.

5. Develop Distributed Sensing

In rapidly changing environments, no leader or central team can maintain sufficient awareness to guide adaptation. Catalytic leaders develop "distributed sensing" capabilities that enable the organization to detect and respond to changes from multiple vantage points.

A telecommunications company exemplified this practice by creating "digital experience scouts", employees across functions who received special training to identify how their digital tools affected customer and employee experience. These scouts provided early warnings of issues and opportunities that would have taken months to surface through traditional reporting channels.

Leadership Skills for Transformation

The shift from command to catalytic leadership requires developing new skills that many leaders haven't cultivated in their careers:

1. Systems Thinking

Traditional leadership emphasizes optimizing parts; transformation requires understanding whole systems. Leaders need to develop:

- Ability to identify interconnections and dependencies
- Recognition of feedback loops and unintended consequences
- Capacity to address root causes rather than symptoms
- Comfort with non-linear change dynamics

2. Productive Discomfort

Transformation inevitably creates discomfort. Effective leaders develop:

- Personal tolerance for ambiguity and uncertainty
- Capacity to distinguish productivity from unproductive discomfort
- Ability to support others through challenging transitions
- Resilience in the face of setbacks and surprises

3. Inquiry Orientation

Complex transformation requires continuous learning. Leaders need:

- Genuine curiosity about diverse perspectives
- Skill in asking powerful questions rather than providing answers
- Ability to surface and test assumptions
- Comfort with not knowing and discovering together

4. Power Intelligence

Catalytic leadership requires sophisticated understanding of how power operates. Leaders must develop:

- Awareness of their own power and how it affects others
- Ability to share power without avoiding responsibility
- Skill in navigating organizational politics constructively
- Capacity to use formal authority to create space for emergence

5. Narrative Crafting

Transformation requires meaning-making amidst change. Leaders need:

- Ability to connect technological change to human purpose
- Skill in creating compelling narratives about the journey
- Capacity to integrate multiple perspectives into shared stories
- Comfort with evolving narratives as transformation unfolds

Leadership Development for Transformation

Organizations serious about transformation must rethink leadership development to build these capabilities. This means moving beyond traditional approaches and being focused on individual competencies to more systemic development that includes:

- Collective leadership development that builds transformation capabilities in intact teams
- Action learning projects that develop skills through real transformation challenges
- Peer learning communities that enable shared sense-making across boundaries
- Immersive experiences that disrupt established mental models and practices
- Reflection disciplines that deepen learning from transformation experiences

A global insurance company exemplified this approach by creating a "Transformation Leadership Lab" where leaders worked in cross-functional teams on real digital initiatives while receiving coaching, peer feedback, and structured reflection opportunities. This integrated development of approach-built transformation capabilities was more effective than traditional leadership programs could have.

From Hero to Host

Perhaps the most fundamental shift in the leadership paradigm is moving from the "leader as hero" to the "leader as host" mindset. The hero leader solves problems, provides answers, and drives change through personal effort. The host leader creates spaces where others can contribute, connect, and create solutions together.

This shift is particularly crucial for HR leaders guiding transformation. Rather than positioning themselves as experts with answers, effective HR transformation leaders function as hosts who:

- Create spaces for diverse stakeholders to shape change
- Connect people with complementary capabilities and perspectives
- Curate resources and learning that enable distributed leadership
- Cultivate conditions where innovation and adaptation flourish

A healthcare organization's CHRO exemplified this approach during their patient experience transformation. Rather than developing a comprehensive plan, she created a "Patient Experience Design Collaborative" where staff from all levels and functions worked with patients to redesign experiences using new digital tools. Her role focused on providing resources, removing barriers, and ensuring psychological safety, hosting the conditions for others to lead transformation.

Leadership Systems, Not Just Leaders

Finally, true transformation requires attending to leadership systems—the structures, processes, and norms that shape how leadership happens—not just individual leaders. These systems include:

1. Governance structures that determine decision rights and accountability
2. Performance metrics that signal what matters in transformation
3. Talent processes that identify and develop transformation leadership
4. Recognition approaches that reinforce desired leadership behaviors
5. Meeting disciplines that model new ways of working together

A retail organization struggling with digital transformation discovered that their leadership system, particularly their quarterly business review process, was reinforced command leadership despite their stated desire for more catalytic approaches. By redesigning this process to emphasize learning rather than judgment, they shifted the entire leadership culture toward more effective transformation support.

The HR Leader's Role

HR leaders occupy a crucial position in shifting the leadership paradigm. Beyond developing catalytic capabilities in themselves, they can:

1. Advise senior leaders on leadership implications of technological change

2. Design talent systems that identify and develop transformation leaders

3. Create leadership experiences that build catalytic capabilities

4. Coach leaders navigating the transition between paradigms

5. Shape cultural norms that reinforce catalytic leadership practices

This advisory role positions HR not as transformation executors but as transformation enablers, professionals who help the organization develop the leadership conditions where technology-enabled change can flourish.

As we turn to the next chapter on process reimagination, remember this essential insight: No technology, however powerful, can overcome leadership paradigms that constrain its potential. True transformation requires not just implementing new tools but evolving how leadership happens throughout the organization.

50

5

Process Reimagination: Beyond Digitization

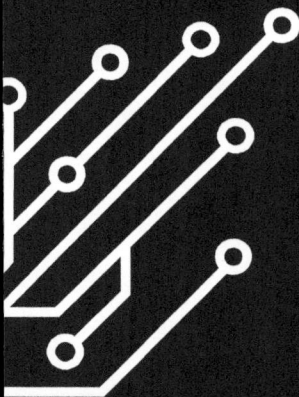

52

CHAPTER 5

Process Reimagination: Beyond Digitization

"We spent two years moving our paper-based performance management process online," the HR director explained, "but we're still not seeing any meaningful improvement in performance or development outcomes."

This scenario is repeated across countless organizations which illustrates a common transformation mistake: digitizing broken processes rather than fundamentally reimagining how work should happen in a technologically enabled environment. The result is what I call "paving the cowpaths", using sophisticated technology to more efficiently do things that shouldn't be done at all.

In this chapter, we'll explore how true transformation requires process reimagination that goes beyond digitization to fundamentally rethink how work happens.

The Digitization Trap

Most organizations fall into the digitization trap, the belief that moving existing processes from analog to digital formats constitutes transformation. This approach produces several predictable problems:

1. Embedded inefficiency: Digitizing flawed processes simply makes them more efficiently flawed

2. Lost opportunity: Building around existing constraints ignores new possibilities enabled by technology

3. Experience degradation: Adding digital layers to analog processes often creates more work rather than less

4. Transformation disillusionment: Disappointing results from digitization create resistance to future initiatives

A healthcare organization I advised had digitized their leave request process, moving from paper forms to online submission. While this created a better audit trail, it increased processing time because the underlying approval workflow—with its five hierarchical sign-offs—remained unchanged. Technology merely created a more visible record of an inherently broken process.

From Digitization to Reimagination

True transformation requires moving beyond digitization to process reimagination that is fundamentally rethinking how work should happen given the capabilities technology now enables. This shift involves three key mindset changes:

1. From Efficiency to Effectiveness

Digitization focuses primarily on efficiency that is doing the same things faster or with fewer resources. Reimagination focuses on effectiveness, which is achieving desired outcomes through potentially very different means.

A financial services firm initially approached their client onboarding transformation as a digitization project, creating online versions of their existing 32-step process. Results were disappointing until they reframed the challenge around effectiveness: "How might we create an onboarding experience that builds trust and understanding while meeting regulatory requirements?"

This effective orientation led them to completely reimagine onboarding as a relationship-building journey supported by technology rather than a document-processing pipeline. The result was 60% faster on-boarding, higher client satisfaction, and better compliance outcomes, A transformation that digitization alone couldn't have delivered.

2. From Process to Experience

Digitization typically focuses on internal processes visible to the organization. Reimagination starts with the human experience of customers, employees, or other stakeholders and works backward to design processes that create desired experiences.

A government agency transformed their citizen service approach by shifting from process-first to experience-first thinking. Rather than digitizing their existing departmental processes, they mapped the citizen journey across services and designed integrated digital experiences that spanned traditional departmental boundaries. This reimagination reduced the average time to complete common citizen needs by 70% while increasing satisfaction and compliance.

3. From Incremental to Zero-Based

Digitization takes existing processes as its starting point and seeks incremental improvements. Reimagination adopts a zero-based approach that asks: "If we were designing this today, with no constraints from the past and full awareness of current technological possibilities, what would we create?"

A manufacturing company exemplified this approach when implementing IoT sensors in their equipment. Rather than simply digitizing their existing maintenance schedules, they asked: "With continuous performance data now available, how should maintenance fundamentally change?" This zero-based thinking led them to reimagine maintenance as a predictive, algorithm-driven practice rather than a calendar-based process which resulted in reducing downtime by 63% and maintenance costs by 41%.

The Reimagination Framework

To move beyond digitization to true process reimagination, I recommend a structured framework with five key phases:

1. Purpose Clarification

Reimagination begins by clearly articulating the fundamental purpose of the work, not how it's currently done, but what outcomes it ultimately seeks to create. This goal emphasis frequently shows that current procedures fulfill outdated needs that might not be relevant today.

A retail organization embarking on workforce scheduling transformation began by clarifying the purpose: "Creating staffing patterns that simultaneously maximize customer service, employee wellbeing, and operational efficiency." This purpose clarification revealed that their existing process overemphasized operational efficiency at the expense of other equally important outcomes.

Key questions in this phase include:

- What human or business need does this work ultimately serve?
- How would we know if we achieved this purpose excellently?
- Which aspects of the current process advance this purpose?
- What historical constraints shaped our current approach that may no longer apply?

2. Experience Design

With clear purpose established, reimagination shifts to designing the ideal experience for all stakeholders involved be it the customers, employees, partners, and others. This phase deliberately delays technology considerations to focus first on the human experience we seek to create.

A hospitality company reimagining their guest check-in process mapped the emotional journey they wanted guests to experience, from feeling welcomed and recognized to feeling oriented and settled before considering any technological solutions. This experience design became the standard against which all process and technology decisions were evaluated.

Key activities in this phase include:

- Mapping current experiences to identify pain points and bright spots
- Articulating principles for ideal stakeholder experiences
- Creating experienced personas that represent different stakeholder needs
- Designing journey maps for ideal future experiences

3. Value Stream Optimization

With desired experiences defined, reimagination examines the end-to-end value stream. All activities required to deliver those experiences to eliminate waste, reduce friction, and maximize value creation.

A financial services organization reimagining their mortgage process identified that 62% of current activities added no value to either the customer experience or risk management. By eliminating these non-value-adding steps before applying technology, they created a fundamentally more effective process rather than merely digitizing waste.

Key questions in this phase include:

- Which activities directly contribute to our purpose and desired experiences?
- Where do handoffs, approvals, or controls add friction without proportional value?
- How might work be restructured to create flow and eliminate batching?
- What assumptions about sequence or dependencies could be challenged?

4. Technology Enablement

Only after clarifying purpose, designing experiences, and optimizing value streams does reimagination turn to technology enablement which is identifying how digital tools can support and enhance the reimagined work.

A manufacturing organization implementing a new product life-cycle management system first redesigned their development process around concurrent engineering principles. Only then did they configure the technology to support this fundamentally different work approach. The result was 40% faster time-to-market compared to competitors who had implemented the same technology on top of sequential development processes.

Key considerations in this phase include:

- How can technology automate routine aspects while enhancing human judgment?
- Where can technology remove distance or time constraints in collaboration?
- How might data and analytics augment decision quality or velocity?
- What new capabilities or experiences does technology make possible?

5. Continuous Evolution

Unlike traditional process design that seeks stability, reimagination establishes mechanisms for continuous evolution such as technology, customer expectations, and business conditions change.

A telecommunications company built quarterly "experience reflection sessions" into their customer service operations, where frontline staff and customers collaboratively evaluated how well digital and human elements were working together and recommended adaptive changes. This discipline enabled their service model to continuously evolve rather than requiring periodic "transformation" initiatives.

Key elements of this phase include:

- Establishing feedback loops from multiple stakeholder perspectives
- Creating regular forums to reflect on experience data
- Empowering frontline teams to implement continuous improvements
- Building mechanisms to capture and scale successful adaptations

The Middle Path: Bimodal Process Design

While zero-based reimagination is the ideal, practical constraints sometimes require a more measured approach. In these cases, I recommend "bimodal process design" that distinguishes between:

- Core transaction processes that benefit from standardization and efficiency
- Value-creating interaction processes that benefit from personalization and judgment

A pharmaceutical company applied this approach to clinical trial management. They standardized and digitized transactional elements like data collection and documentation while reimagining interaction elements like patient communication and investigator collaboration. This bimodal approach delivered efficiency where it mattered without sacrificing the human elements critical to trial success.

Common Reimagination Patterns

Across industries and functions, several common patterns emerge when organizations successfully reimagine processes for the digital era:

1. From Batch to Flow

Traditional processes batch work to optimize resource utilization. Re-imagined processes leverage technology to create continuous flow that optimizes cycle time and responsiveness.

A government agency transformed their benefits processing from monthly batch cycles to continuous processing enabled by automated eligibility verification. This shift reduced average waiting time from 17 days to 36 hours while improving accuracy through real-time data integration.

2. From Sequential to Concurrent

Traditional processes organize work sequentially to manage dependencies. Reimagined processes leverage collaborative technologies to enable concurrent work that dramatically accelerates outcomes.

An engineering firm reimagined their design review process from sequential functional sign-offs to concurrent collaborative reviews using digital prototypes and annotation tools. This shift reduced review cycles from weeks to days while improving design quality through integrated perspective.

3. From Centralized to Distributed

Traditional processes centralize control to ensure consistency. Reimagined processes distribute capability to the edge while maintaining central visibility and learning.

A retail bank reimagined lending from a centralized approval process to a distributed model where frontline staff used algorithm-supported decision tools within guardrails. This shift reduced approval time from days to minutes while improving decision quality through data-augmented human judgment.

4. From Standardized to Personalized

Traditional processes standardize experience for efficiency. Reimagined processes leverage data to create personalized experiences on a scale.

An educational institution reimagined their student support from standardized check-ins to a personalized outreach model using predictive analytics to identify student-specific needs. This approach improved retention by 23% while reducing total support staff time through more targeted interventions.

5. From Reactive to Predictive

Traditional processes respond to events after they occur. Reimagined processes leverage data to predict and prevent issues before they happen.

A manufacturing organization transformed maintenance from a reactive "fix-it-when-broken" approach to a predictive model using sensor data and machine learning. This reimagination reduced unplanned downtime by 78% and extended equipment life through earlier intervention.

Reimagination in HR

HR functions face unique challenges in process reimagination due to their complex stakeholder landscape and the hybrid nature of transactional and relational tasks. Successful HR transformations apply several key principles:

1. Employee Experience First

Rather than organizing around HR specialties, HR processes reimagined key employee experience moments, from joining to developing to transitioning.

A technology company redesigned their entire HR operating model around employee journeys instead of traditional HR functions. This shift integrated previously siloed processes into a seamless experience while clarifying where high-touch human support adds the most value.

2. Self-Service Plus

Reimagined HR processes distinguish between transactions that are best suited for self-service and those where human guidance adds critical value.

A healthcare organization implementing a new HRIS reimagined their approach to employee life events. They created an intuitive self-service system for transactional elements while establishing "life guides" who proactively supported employees during significant transitions, providing personalized assistance beyond what technology alone could offer.

3. Decision Intelligence

Rather than just reporting data, reimagined HR processes embed analytics directly into decision points to enhance human judgment.

A manufacturing company transformed their workforce planning by integrating predictive modeling directly into manager tools instead of producing separate reports. This approach improved decision quality by making data insights available precisely when and where they were needed.

4. Community Enablement

Beyond managing formal programs, reimagined HR processes build infrastructure for peer-to-peer connection and learning.

A professional services firm reimagined talent development by shifting from a centrally managed training catalog to a learning ecosystem that combined curated resources, collaboration spaces, peer mentoring, and practice communities. This transformation increased participation in development activities while reducing central L&D costs.

The Reimagination Process

Successfully moving from digitization to reimagination requires a structured approach that overcomes organizational inertia. Key strategies include:

1. Start with Pain

Rather than attempting comprehensive reimagination, start with processes that cause the most stakeholder pain or create the biggest strategic constraints.

A financial services organization began their transformation journey by reimagining their client onboarding process, the process responsible for their lowest customer satisfaction scores and highest drop-off rates. This targeted focus created immediate, visible impact, building momentum for broader reimagination.

2. Involve the Edges

Include not just those who perform the process, but also upstream suppliers and downstream recipients of process outputs.

A manufacturing company redesigning their production planning engaged suppliers, production teams, and sales representatives in a collaborative workshop. This broader involvement revealed integration opportunities that process specialists alone would have missed.

3. Make it Tangible

Use visualization, simulation, and prototyping to bring reimagined processes to life before full-scale implementation.

A healthcare organization created "experience prototypes" for their reimagined patient scheduling process using simple technology mock-ups and role-play. This hands-on approach helped stakeholders evaluate the human impact of proposed changes before technical development began.

4. Run Parallel Experiments

Rather than a single redesign, develop multiple competing reimagination concepts and test them in parallel.

A retail organization testing a new inventory management approach ran parallel experiments with three different models in comparable stores. This strategy provided comparative data that informed their final design while enhancing broader organizational learning about digitally enabled inventory practices.

Reimagination Leadership from HR

HR leaders are uniquely positioned to drive process reimagination due to their enterprise-wide perspective and expertise in human experience. Their role includes:

1. Advocating for experience-centred design that integrates technological capability with human needs
2. Facilitating reimagination efforts to help functional teams move beyond digitization
3. Identifying cross-functional opportunities that break down traditional process silos
4. Sharing lessons and best practices to accelerate learning across transformation initiatives
5. Developing reimagination skills through training programs and curated experiences

When HR leaders embrace this role, they shift from being process owners to becoming transformation enablers, helping organizations build the capabilities necessary for ongoing adaptation.

As we move into the next chapter on skills and capabilities, remember: True transformation doesn't come from doing the same things differently but from doing fundamentally different things. Process reimagination, moving beyond digitization to rethink how work happens—is essential for technology to achieve its full transformative potential.

6

Skills for the Future: Building Transformative Capabilities

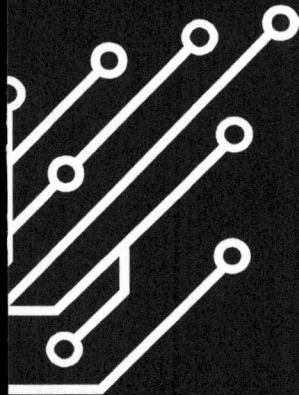

CHAPTER 6

Skills for the Future: Building Transformative Capabilities

"We've invested millions in new technology," the COO lamented, "but we don't have the skills to leverage it. Our people are overwhelmed trying to figure out how to work in this new digital environment."

This scenario—common across industries and functions—highlights a critical aspect of transformation that organizations often underestimate: the capability gap between existing skills and those needed to unlock technology's full potential.

In this chapter, we'll explore why true transformation requires deliberate capability development at the individual, team, and organizational levels—and how HR can lead this crucial shift.

The Transformation Capability Gap

Most organizations begin transformation with a technology-first mindset, underestimating the capability development required. This leads to what I call the "transformation capability gap", the gap between:

- What technology can do (its technical potential)
- What people know how to do with it (their practical capability)

This gap explains why organizations with identical technologies see drastically different results. The true differentiator isn't the technology itself but the human and organizational capabilities that support it.

A manufacturing company implemented advanced analytics tools in two similar plants. Six months later, one plant achieved a 17% productivity improvement, while the other showed little change. The key difference? One plant had invested in building analytical thinking skills, creating cross-functional problem-solving teams, and establishing learning routines to translate insights into action.

From Training to Capability Building

Traditional approaches to closing skill gaps focus on training—teaching knowledge and developing technical proficiency. While important, training alone is insufficient for transformation. True change requires broader capability building, including:

1. **Technical skills** – Proficiency with specific tools and technologies

2. **Adaptive capabilities** – The ability to learn, problem-solve, and navigate ambiguity

3. **Social capacities** – Collaboration, influence, and network-building

4. **Mindset evolution** – Shifting underlying beliefs, values, and identity elements

A retail organization rolling out in-store analytics initially focused only on technical training—teaching employees how to operate new dashboards. However, meaningful results emerged only when they expanded their approach to include:

- Collaborative problem-solving using data (adaptive)
- Cross-functional solution implementation (social)
- Shifting from intuition-based to data-informed decision-making (mindset)

This holistic strategy produced not just technical proficiency but the transformative capability to fundamentally change how work was done.

The Critical Capabilities Matrix

To develop strategic capabilities, organizations can use the Critical Capabilities Matrix—a tool that maps capability needs across three categories and three organizational levels.

Capability Categories:

1. **Technical Capabilities** – Skills for technology usage and application

2. **Business Capabilities** – Skills for applying technology to create business value

3. **Transformative Capabilities** – Skills for leading and navigating change

Organizational Levels:

1. **Individual Contributors** – Employees using technology in daily work

2. **Team Leaders** – Managers guiding others in technology application

3. **Senior Leaders** – Executives shaping transformation strategy

For each intersection, organizations can identify critical capabilities and build a targeted development roadmap.

A financial services firm applied this matrix to guide their digital transformation efforts. At the intersection of "Transformative Capabilities" and "Team Leaders," they identified "Leading through uncertainty" as a priority skill. To develop it, they designed a program that combined scenario planning, adaptive leadership practices, and peer coaching.

The Learning Ecosystem Approach

Traditional learning approaches—centralized, episodic, and content-heavy—fail to support the dynamic, contextual capability-building that transformation demands. Instead, successful organizations develop a "learning ecosystem" that fosters continuous capability growth through multiple integrated channels:

1. **Formal learning** – Structured programs for foundational knowledge and skills

2. **Social learning** – Communities, networks, and collaborative practice

3. **Experiential learning** – Project assignments, rotations, and stretch opportunities

4. **Performance support** – Tools, resources, and coaching embedded in workflow

5. **Self-directed learning** – Curated resources for individual exploration

A healthcare organization adopting telehealth implemented a comprehensive learning ecosystem rather than just training clinicians on the new platform. Their approach included:

- Structured telehealth certification (formal)

- Practice communities where clinicians shared experiences (social)

- Progressive telehealth cases of increasing complexity (experiential)

- Virtual coaches available during initial consultations (performance support)

- Curated research on telehealth best practices (self-directed)

This ecosystem-driven approach built deeper capability more effectively than traditional training alone.

Adaptive Learning Journeys

Rather than one-size-fits-all programs, transformation requires adaptive learning journeys that account for different starting points, learning styles, and application contexts.

A government agency transitioning to digital citizen services designed multiple learning pathways that:

- Adjusted to different roles and levels of digital fluency
- Offered flexibility in learning modalities (in-person, virtual, self-paced)
- Provided tailored application support for specific service contexts
- Created personalized stretch challenges based on individual progress
- Included remediation options for those needing additional support

This personalized strategy optimized capability development by focusing resources where they were most needed while maintaining a cohesive overall framework.

The 70-20-10 Reality

Research consistently shows that adult learning follows a 70-20-10 distribution:

- 70% from challenging experiences and practice
- 20% from developmental relationships and feedback
- 10% from formal instruction and content

Yet, most transformation initiatives invert this ratio, overemphasizing formal training. Successful organizations align their capability-building efforts with this reality by:

1. Creating structured experiential opportunities to develop critical skills
2. Establishing coaching and peer-learning networks to accelerate development
3. Using formal learning as preparation for and reinforcement of experiential learning

A manufacturing company enhanced its data literacy through project-based learning, where cross-functional teams applied analytical tools to solve real business challenges. Formal training accounted for only 10% of the learning experience, while most learning occurred through hands-on application and coaching from data scientists, who acted as "analytical translators" for the teams.

Critical Mindset Shifts

Beyond technical skills, transformation requires fundamental shifts in how people approach work, perceive value, and define their roles. These mindset changes are often the most challenging but also the most transformative.

Key Mindset Shifts for Digital Transformation:

1. **From stability to adaptability** – Seeing change as the norm, not the exception

2. **From knowing to learning** – Valuing curiosity and growth over expertise

3. **From control to influence** – Leading through networks rather than hierarchy

4. **From process to outcome** – Prioritizing results over rigid procedures

5. **From specialization to integration** – Recognizing connections across disciplines

A utility company implementing grid optimization technology found that the biggest barrier wasn't technical proficiency but mindset, specifically, shifting from a strict safety culture to one that balanced safety with adaptive problem-solving. They successfully drove this change through storytelling, peer modelling, and creating safe spaces to practice new ways of thinking.

Team Capability: The Missing Middle

Most capability development focuses either on individuals or on organization-wide initiatives, neglecting the critical middle layer of team capability. Yet teams, not individuals or organizations— are the primary units where technology is applied to create value.

Essential team capabilities for transformation include:

1. Collective sensemaking: Interpreting and making sense of complex information together

2. Distributed cognition: Leveraging diverse knowledge, expertise and perspectives

3. Adaptive coordination: Adjusting roles and workflows dynamically as conditions evolve

4. Collaborative learning: Developing shared knowledge through practice and reflection

5. Boundary spanning: Effectively connecting across organizational silos and functions

A healthcare organization adopting a new electronic health record prioritized team capability development alongside individual skill-building. They established team learning labs where entire clinical units practiced using the system together, developing team-specific protocols and mutual support strategies. These teams significantly outperformed departments that focused solely on individual proficiency.

Learning Integration Mechanisms

To ensure that capability-building leads to performance improvement, organizations need specific mechanisms to integrate learning into daily work. These mechanisms include:

- Application assignments that require employees to apply new skills in real work scenarios

- Action learning projects that address business challenges while building capabilities

- Peer coaching circles that offer support during the skill application phase
- Knowledge exchanges that capture and distribute learning across teams
- Performance dialogues that link capability development directly to outcomes

A financial services company incorporated these mechanisms into their digital transformation, introducing bi-weekly "Digital Breakthrough Sessions." In these sessions, teams shared concrete examples of how they applied new capabilities to improve customer outcomes. This approach fostered accountability for implementation while accelerating cross-team learning.

The Skill Acquisition Curve

Organizations often underestimate the time it takes for new capabilities to translate into performance improvement. This leads to what I call the "implementation dip"—a temporary drop in performance as employees transition from unconscious competence with old approaches to conscious incompetence with new ones.

How Successful Organizations Manage the Implementation Dip:

1. Setting realistic timelines for capability development and performance improvement

2. Providing enhanced support during the transition phase

3. Recognizing and celebrating learning and progress, not just performance outcomes

4. Temporarily adjusting performance expectations to accommodate the learning curve

5. Communicating the skill acquisition curve model to normalize the implementation dip

A retail organization introducing new customer analytics tools acknowledged this curve upfront, telling store managers:

"We expect your performance metrics may initially decline for the first 6–8 weeks as you learn the new system. This temporary dip is normal, and those who persist through this phase consistently outperform within three months."

This clear and realistic framing prevented teams from prematurely abandoning new capabilities during the inevitable early-stage performance dip.

The Capability Development Portfolio

Just as organizations manage investment portfolios with varying time horizons, they should develop capability portfolios that balance:

1. Now capabilities: Skills required for immediate performance with current technology

2. Next capabilities: Skills required for planned technology implementations

3. Future capabilities: Foundational skills for anticipated but undefined future needs

A manufacturing organization maintained this balanced capability portfolio by strategically allocating development resources across three horizons:

- 60% to current digital manufacturing skills

- 30% to capabilities needed for an upcoming IoT implementation

- 10% to foundational skills like systems thinking and data literacy, ensuring readiness for future advancements

This approach helped them avoid the common mistake of focusing solely on immediate skill gaps while neglecting the capabilities necessary for long-term transformation.

HR's Transformative Role in Capability Building

HR leaders must transition from learning providers to capability architects who:

1. Identify and map critical capabilities required across transformation initiatives
2. Design integrated learning ecosystems rather than one-off training programs
3. Embed learning into daily work through deliberate integration mechanisms
4. Curate and amplify emerging best practices discovered through transformation
5. Develop organizational capacity to continuously build and adapt skills

This shift positions HR not just as a service provider but as a strategic enabler of digital transformation.

A global professional services firm embodied this shift by creating a "Transformation Capability Lab" within HR. This dedicated function worked across digital initiatives to:

- Identify recurring capability needs
- Develop scalable learning solutions
- Accelerate cross-initiative knowledge sharing

This centralized approach significantly improved both the efficiency and effectiveness of capability development compared to fragmented, initiative-specific training efforts.

Measuring Capability Development

Organizations need more sophisticated ways to measure capability development, moving beyond traditional learning metrics like completion rates and satisfaction scores.

More Effective Capability Metrics:

1. **Capability assessments** – Measure actual skill application, not just knowledge retention
2. **Performance correlations** – Link capability development to business outcomes
3. **Acceleration metrics** – Track improvements in time-to-competence
4. **Adaptation indicators** – Assess how capabilities evolve with changing demands
5. **Diffusion measures** – Analyse how new capabilities spread across teams and functions

A telecommunications company adopted an advanced capability measurement approach for their digital transformation. Their system included:

- Quarterly capability assessments
- Targeted experiments to correlate skills with performance improvements
- Network analysis to track how capabilities spread across organizational boundaries

This evidence-based approach allowed them to continuously refine their capability-building strategy based on real impact rather than assumptions.

The Long View on Capability Building

Perhaps most importantly, organizations must recognize that transformative capability building is not a one-time event but an ongoing discipline. The capabilities needed today will continue to evolve as technology advances and competitive landscapes shift.

Organizations that sustain long-term transformation develop what I call "second-order capabilities"—meta-skills that enable continuous adaptation:

1. **Learning agility** – The ability to rapidly acquire new skills and knowledge

2. **Pattern recognition** – The capacity to anticipate emerging capability requirements

3. **Knowledge creation** – The ability to generate new insights through practice

4. **Social learning** – Learning through collaboration and networks

5. **Reflection discipline** – Extracting insights from experience through deliberate reflection

A pharmaceutical company implementing AI in drug discovery recognized that technical skill requirements would constantly evolve. Rather than chasing moving targets, they focused on developing these second-order capabilities through structured practices such as learning retrospectives, cross-functional communities of practice, and rotational experiences. This approach helped them build adaptive capacity alongside current technical skills.

As we move to the next chapter on change architecture, remember this essential truth: Technology provides possibility, but capability determines whether that possibility leads to transformation. When we invest as thoughtfully in human and organizational capability as we do in technological capability, we create the conditions for sustained transformative change.

7

The Change Architecture: Designing for Adaptation

CHAPTER 7

The Change Architecture: Designing for Adaptation

A transformation initiative appeared to have everything needed for success—executive sponsorship, substantial funding, cutting-edge technology, dedicated project teams, and detailed implementation plans. Yet, eighteen months later, despite technical deployment being "on time and on budget," actual changes in work processes were minimal. The promised transformation remained elusive.

This scenario—painfully familiar to many organizations—highlights a fundamental truth: Implementing technology is necessary but insufficient for transformation. True transformation requires what I call "change architecture"—the deliberate design of conditions, processes, and structures that enable adaptation at individual, team, and organizational levels.

In this chapter, we'll explore how HR leaders can move beyond traditional change management to become architects of sustainable transformation.

From Change Management to Change Architecture

Most organizations approach transformation through the lens of "change management"—a discipline focused on managing transitions from current to future states. While valuable, this approach has significant limitations in complex, technology-enabled transformations:

1. Assumes a linear progression from the current state to a predefined future state

2. Emphasizes execution of pre-determined plans over adaptive learning

3. Views resistance as an obstacle rather than a source of valuable feedback

4. Treats change as an event rather than an ongoing evolution

Change architecture takes a more sophisticated approach that:

1. Creates conditions for continuous adaptation rather than discrete transitions

2. Establishes enabling structures that support emergent learning

3. Designs feedback loops that accelerate adjustments and innovation

4. Builds participatory processes that engage the organization's collective intelligence

A healthcare organization implementing precision medicine technologies demonstrated this shift in approach. Instead of developing a rigid change management plan with fixed adoption targets, they established a "learning acceleration infrastructure" that included:

- Clinical innovation teams
- Cross-functional reflection forums
- Data feedback systems
- Flexible resource allocation

This allowed the organization to continuously refine how the technology could best improve patient outcomes.

The Four Dimensions of Change Architecture

Effective change architecture addresses four interdependent dimensions, creating the conditions for successful transformation.

1. Direction Architecture

This dimension ensures a shared understanding of transformation goals without over-specifying implementation details. Key elements include:

- **Purpose narrative:** A compelling story that connects technological change to meaningful human and organizational impact

- **Boundary conditions:** Clear non-negotiables that ensure alignment while allowing for local adaptation

- **Success measures:** Balanced indicators that go beyond technical implementation to reflect business and human outcomes

- **Directional principles:** Values-based guidance that informs decision-making

A retail organization implementing customer experience technologies developed a strong purpose narrative focused on "creating moments of delight in every customer interaction" rather than technical adoption targets. Their approach included:

- **Boundary conditions** (ensuring solutions worked for their least digitally savvy customers)

- **Success measures** (spanning customer, employee, and business outcomes)

- **Decision principles** (to guide adaptation at local levels)

2. Learning Architecture

This dimension ensures rapid learning, continuous adjustment, and knowledge sharing. Key elements include:

- **Feedback systems** – Mechanisms to gather, analyze, and act on insights

- **Reflection routines** – Structured debriefs that extract lessons from experience

- **Knowledge networks** – Cross-functional connections that spread learning

- **Experimental structures** – Safe spaces to test and refine new approaches

A manufacturing company implementing IoT technology built a learning architecture that included:

- Monthly sensing forums where frontline teams shared real-time challenges

- Quarterly reflection summits that extracted patterns across locations

- A digital knowledge common to document and refine solutions

- Innovation sandboxes for low-risk experimentation

3. Engagement Architecture

This dimension harnesses collective intelligence while building ownership. Key elements include:

- **Participation platforms** – Broad opportunities for employees to contribute

- **Agency gradients** – Clarification of decision rights at different levels

- **Connection mechanisms** – Bridging formal initiatives with informal networks

- **Meaningful roles** – Enabling employees at all levels to shape transformation

A financial services organization implementing wealth management technologies designed an engagement architecture featuring:

- A representative advisory council shaping overall direction
- Local adaptation teams with decision-making authority
- Technology champions embedded within each team
- Open feedback channels to surface insights and concerns.

4. Support Architecture

This dimension ensures teams have the resources and structures necessary for successful adaptation. Key elements include:

- **Capacity allocation** – Dedicated time for transformation work
- **Skill development** – Just-in-time learning aligned with transformation needs
- **Transition assistance** – Psychological and practical support for employees
- **Resource flexibility** – Dynamic allocation of funds and personnel

A pharmaceutical company implementing digital clinical trials created a robust support architecture that included:

- Protected time for teams to redesign workflows
- Embedded digital coaches providing contextual skill development
- "Transformation dialogue guides" for leaders to support employees
- Rapid-access funds for agile implementation

The Systemic View of Change

Traditional change approaches often focus on isolated interventions—such as communication plans, training programs, or incentives—

without considering how these elements interact within the broader system. Effective change architecture ensures that these components reinforce each other, creating an environment where transformation can truly take root.

Changing architecture takes a systems view, recognizing that various organizational elements must evolve together for transformation to succeed:

1. **Strategic elements:** Purpose, priorities, and success definitions
2. **Structural elements:** Reporting relationships, decision rights, and coordination mechanisms
3. **Process elements:** Workflows, information flows, and resource flows
4. **People elements:** Skills, mindsets, and relationships
5. **Measurement elements:** What gets tracked, how it's evaluated, and the consequences
6. **Cultural elements:** Norms, values, and shared beliefs

A telecommunications company implementing customer experience technologies initially focused solely on training frontline staff, but the results were disappointing. When they adopted a systems approach, realigning performance metrics to prioritize experience over efficiency, adjusting supervisor roles to emphasize coaching, redesigning workflows to support new capabilities, and shifting cultural norms around customer interaction, adoption accelerated dramatically.

This systemic perspective prevents the common transformation mistake of changing one element without adjusting interdependent elements, which often creates unsustainable conflicts within the organization.

From Resistance to Co-Creation

Traditional change management treats resistance as an obstacle, to be overcome through persuasion or mandates. Change architecture, however, reframes resistance as valuable intelligence that can strengthen transformation through a co-creation approach.

Key practices in this approach include:

1. **Legitimate concerns dialogues:** Structured conversations that surface and address substantive issues
2. **Solution co-creation:** Collaborative processes where those affected by change help design implementation
3. **Adaptation ownership:** Clear decision rights for local teams to adjust implementation for their context
4. **Interest integration:** Explicit efforts to align transformation approaches with stakeholder interests

A manufacturing organization implementing advanced analytics faced significant resistance from production supervisors. Instead of forcing compliance, they formed solution teams where supervisors worked alongside data scientists to design analytics applications that addressed their most pressing operational challenges. This co-creation approach not only reduced resistance but also led to better solutions, informed by deep operational knowledge.

The Transformation Roadmap Reimagined

Traditional transformation roadmaps outline predetermined implementation steps in a linear sequence. While this provides clarity, it struggles to accommodate the learning and adaptation required for complex change.

Change architecture instead utilizes "adaptive roadmaps"—providing direction while allowing for evolution:

1. **Fixed elements:** Clear articulation of non-negotiable aspects

2. **Learning milestones:** Reflection points where course adjustments may be needed

3. **Decision gates:** Criteria for proceeding or pivoting based on new insights

4. **Parallel options:** Simultaneous testing of multiple approaches to identify optimal solutions

5. **Emergence zones:** Areas intentionally left undefined, allowing for co-creation during implementation

A healthcare organization implementing telemedicine followed this adaptive roadmap approach. They:

- Fixed regulatory compliance requirements and core platform functionality as non-negotiables.

- Created quarterly learning milestones to adjust strategy based on patient and clinician feedback.

- Defined success criteria for expanding from pilot to full implementation.

- Tested three different workflow models in parallel to identify the most effective one.

- Allowed clinical teams to design specialty-based protocols within emergence zones.

This approach provided sufficient structure for coordinated action while ensuring flexibility for learning and adaptation.

Leadership Throughout the System

Traditional change approaches focus heavily on top-down sponsorship. While executive support remains essential, change architecture distributes leadership throughout the system through what I call the "leadership lattice", an interconnected network of formal and informal leaders who drive transformation at multiple levels:

1. **Executive sponsors:** Senior leaders who provide resources, remove barriers, and connect transformation to strategy
2. **Transformation guides:** Specialists who design change architecture and coach others on transformation practices
3. **Local champions:** Respected peers who model new ways of working and support colleagues
4. **Integration nodes:** Boundary-spanners who connect initiatives and spread learning
5. **Practice pioneers:** Early adopters who experiment with new possibilities and share discoveries

A retail organization implementing new store technologies activated this leadership lattice by:

- Having executive sponsors articulate the transformation's purpose and allocate resources.
- Deploying transformation guides to design learning systems and capability-building initiatives.
- Appointing store-level champions to provide peer coaching on new tools.
- Establishing cross-functional teams to integrate online and in-store experiences.
- Encouraging designated associates to experiment with creative applications and share successful practices.

This distributed leadership model allowed the organization to address transformation challenges at the right level, reducing the burden on senior leadership while accelerating adaptation.

Change Architecture in HR Transformation

HR functions face unique challenges—they must **transform their own operations while supporting broader organizational change.** Effective HR transformations apply **several architectural principles:**

1. The Three-Horizon Approach

Rather than attempting a comprehensive transformation all at once, successful HR functions phase their transformation using a three-horizon approach:

- **Horizon 1:** Stabilize and optimize core transactional services
- **Horizon 2:** Enhance strategic capabilities in targeted areas
- **Horizon 3:** Reimagine the HR function's purpose and operating model

This phased approach allows HR to meet current service expectations while progressively building toward deeper transformation.

2. The Experience-Back Design

Instead of designing transformation from HR process categories, successful HR transformations start with stakeholder experiences and work backward to determine technological and structural requirements.

A global manufacturer structured their HR transformation around employee journeys—such as joining, developing, changing roles, and transitioning—rather than traditional HR process categories. This experience-centred approach guided technology selection, capability building, and role design decisions.

3. The Bimodal Structure

During transformation, HR must simultaneously maintain service excellence while developing new capabilities. This often requires a bimodal structure that distinguishes between:

- Operational excellence mode – Focused on reliability and efficiency in core services
- Transformation mode – Focused on innovation and learning in emerging capabilities

A healthcare organization established a dedicated HR transformation team that could experiment with new approaches, while the core HR function-maintained service delivery. This separation created space for innovation without disrupting operational stability.

Measuring Transformation Progress

Traditional change metrics focus primarily on implementation milestones and adoption rates. While important, these measures provide limited insight into whether true transformation is occurring.

Change architecture requires more sophisticated measurements across four dimensions:

1. **Technical implementation:** Are we deploying the technology as planned?

2. **Capability development:** Are people building the skills to use the technology effectively?

3. **Work evolution:** Are actual work patterns and decisions changing meaningfully?

4. **Outcome realization:** Are we achieving the human and business outcomes we seek?

A financial services organization implemented this measurement approach for their digital advisor platform transformation. Beyond tracking technical deployment and usage metrics, they assessed advisors' confidence in data-informed client conversations, observed how client interaction patterns were evolving, and measured changes in client satisfaction and investment outcomes.

This balanced measurement approach provided a more accurate picture of transformation progress while highlighting specific areas requiring additional support.

The Transformation Office Reimagined

Many organizations establish transformation or project management offices to coordinate change efforts. While valuable for alignment, these functions often become compliance-oriented administrators rather than enablers of adaptation.

Change architecture reimagines this function as a Transformation Enablement Team that:

1. Designs and maintains the change architecture across all four dimensions
2. Facilitates learning and adaptation rather than merely tracking compliance
3. Connects distributed change activities to enable pattern recognition
4. Builds transformation capability throughout the organization
5. Evolves approaches based on emerging needs and discoveries

A global manufacturer exemplified this approach by establishing a transformation team explicitly chartered to "create the conditions for successful change rather than drive implementation from the center." This team spent 70% of its time designing learning systems, building change capability, and enabling cross-boundary integration—activities that accelerated adaptation far more effectively than traditional project management.

From Programs to Platforms

Finally, change architecture shifts from viewing transformation as a collection of time-bound programs to establishing ongoing platforms that enable continuous adaptation. These platforms include:

1. **Learning infrastructures** that continuously capture and share emerging insights
2. **Collaboration architectures** that enable cross-boundary problem-solving
3. **Resource reallocation mechanisms** that shift support toward emerging priorities
4. **Capability development systems** that continuously build needed skills
5. **Distributed experimentation structures** that test and refine new possibilities

A professional services firm exemplified this shift by establishing what they called "transformation platforms" rather than programs. These platforms—including a client solution lab, an internal capability accelerator, and a digital ethics forum—provided ongoing structures for continuous adaptation rather than time-bound implementation projects.

This platform approach recognized that true transformation isn't a destination reached through programs, but an ongoing journey enabled by adaptive infrastructure.

HR as Transformation Architects

For HR leaders, the evolution from change managers to change architects represents a significant opportunity to elevate their strategic contribution. This role involves:

1. Designing holistic change architectures that address all four dimensions
2. Advising leaders on creating conditions for successful adaptation
3. Integrating initiatives to create coherent transformation experiences

4. Building organizational muscles for continuous change

5. Modelling adaptive approaches within HR's own transformation

When HR embraces this architect role, it positions itself not as a transformation support function but as a strategic partner that shapes how the organization approaches and experiences change.

As we move to the next chapter on data and intelligence, remember this essential insight: True transformation requires not just implementing new technologies but creating the conditions where technology can catalyse meaningful adaptation. By designing sophisticated change architectures, HR leaders enable the organization to continuously evolve in ways that technology alone cannot make possible.

8

Data and Decisions:
The Intelligence Layer

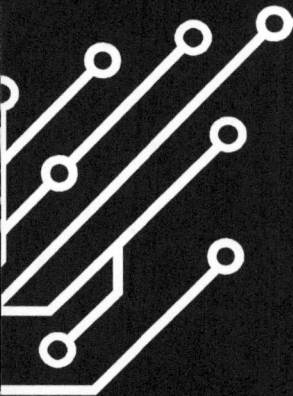

CHAPTER 8

Data and Decisions: The Intelligence Layer

The dashboard was impressive, it displayed real-time visualizations of workforce metrics that would have been impossible to compile manually. Yet, the executive team continued making decisions exactly as they had before the system was implemented. The data was available but remained unused, a sophisticated digital layer disconnected from actual decision-making.

This scenario illustrates a common transformation challenge: organizations invest heavily in data infrastructure without equivalent attention to the human and organizational elements required to translate data into intelligence and intelligence into action. The result is what I call "data-rich, insight-poor" organizations—those that possess information but lack the capability to use it to transform.

In this chapter, we'll explore how true transformation requires developing an effective intelligence layer—the human and technological capabilities that connect data to decisions in ways that fundamentally improve outcomes.

From Data to Intelligence

Organizations often conflate data with intelligence, if simply having access to information will automatically improve decision-making. This assumption ignores the critical transformations that must occur:

Data → Information → Insight → Intelligence → Action

Each transition in this chain requires specific capabilities:

1. **Data to Information:** Organizing raw data into meaningful patterns and relationships

2. **Information to Insight:** Interpreting these patterns to identify implications and opportunities

3. **Insight to Intelligence:** Combining insights with context and judgment to inform decisions

4. **Intelligence to Action:** Translating intelligence into changed behaviors and decisions

A manufacturing organization implemented comprehensive operational dashboards but saw minimal impact until they built capabilities at each transition point: data scientists who transformed raw sensor data into meaningful production metrics, analysts who identified performance patterns and their causes, leaders skilled at contextualizing these insights within business priorities, and operational processes redesigned to incorporate data-informed decisions.

This systematic approach to building the complete intelligence chain, not just the data infrastructure, enabled technology to transform how work happened, rather than simply digitizing information delivery.

The Decision Architecture

Most data initiatives focus on supply—generating, organizing, and visualizing information. Transformative approaches balance this with equal attention to demand understanding of how decisions happen and designing intelligence systems to fit decision realities.

This requires mapping what I call the "decision architecture", the patterns of how, when, where, and by whom different types of decisions are made. Key elements include:

1. **Decision typology:** Categories of decisions with different characteristics

2. **Decision rights:** Who has authority for different decision types

3. **Decision processes:** How information flows into deliberation and choice

4. **Decision supports:** Tools, forums, and resources that enable effective decisions

5. **Decision rhythms:** When different decisions occur and their cadence

A healthcare organization transformed their approach to staffing decisions by mapping this architecture. They discovered that unit managers made daily adjustments based on immediate patient needs, department heads made weekly allocations based on an anticipated case mix, and executives made quarterly investments in new positions based on strategic priorities. By designing differentiated intelligence approaches for each decision type, they dramatically improved both efficiency and effectiveness compared to their previous one-size-fits-all staffing dashboard.

Human-Centered Intelligence Design

Technology-first approaches to intelligence often create sophisticated systems that fail to account for human cognitive realities. Transformative approaches apply human-centered design that considers:

1. **Cognitive load:** How much information humans can effectively process

2. **Decision biases:** Systematic patterns that affect human judgment

3. **Attention dynamics:** How focus shifts in different contexts and roles

4. **Collaborative sensemaking:** How groups interpret information together

5. **Trust factors:** What enables or inhibits confidence in data-driven insights

A financial services organization redesigned their risk intelligence approach based on these human factors. Rather than comprehensive risk dashboards that overwhelmed managers with information, they created focused risk alerts highlighting specific anomalies requiring attention, embedded contextual guidance that helped counter common decision biases, and established collaborative review protocols that improved collective judgment about complex risk situations.

This human-centered approach improved both the quality and consistency of risk decisions by designing how people process and use information rather than maximizing data delivery.

The Augmentation Mindset

Organizations often approach intelligence either as automation that replaces human judgment or as passive information that informs unchanged decision processes. Transformative approaches adopt what I call the "augmentation mindset"—deliberately designing intelligence systems that enhance human capabilities in complementary ways.

This mindset recognizes that:

- Algorithms excel at pattern recognition across large datasets.
- Humans excel at contextual interpretation and creative response.
- Neither alone achieves what both can accomplish together.

A retail organization exemplified this approach in merchandise planning. Rather than either automating decisions or simply providing data to unchanged processes, they designed a collaborative intelligence system where algorithms identified sales patterns and suggested initial assortment plans, human merchants applied contextual knowledge about brand strategy and emerging trends, and the combined plan underwent algorithmic simulation testing before implementation.

This augmentation approach consistently outperformed both purely algorithmic and purely human approaches by leveraging the complementary strengths of each.

The Intelligence Ecosystem

Organizations often implement intelligence capabilities in fragmented ways—HR analytics separate from operational analytics separate from customer analytics. This fragmentation creates partial views that miss important connections.

Transformative approaches create what I call an "intelligence ecosystem", interconnected capabilities that enable more holistic understanding and decision-making. Key ecosystem elements include:

1. **Cross-domain data integration:** Connecting information across traditional silos

2. **Multi-level intelligence views:** Enabling exploration from enterprise to individual levels

3. **Balanced indicator sets:** Metrics spanning different time horizons and value dimensions

4. **Causal mapping capabilities:** Tools for understanding relationships between factors

5. **Scenario modeling tools:** Methods for exploring potential futures and options

A manufacturing organization created this ecosystem approach for their production facilities. Their intelligence platform integrated equipment performance, quality metrics, workforce data, and customer satisfaction indicators into a connected view that enabled leaders to understand how these factors interacted. This ecosystem revealed previously invisible relationships—such as how specific team composition factors affected quality for certain product types—that created entirely new improvement opportunities.

Decision Support at Scale

Most intelligence initiatives focus primarily on supporting senior leadership decisions. While important, this narrow focus overlooks the transformative potential of improving the thousands of daily operational decisions across the organization.

Transformative approaches create multi-layered decision support that:

1. **Enables strategic decisions** at the executive level
2. **Supports tactical decisions** by middle management
3. **Improves operational decisions** at the frontline level

A telecommunications company exemplified this approach in their network management transformation. They created integrated intelligence systems that provided executives with investment prioritization insights for infrastructure development, gave regional managers recommendations for optimizing network resources, and equipped field technicians with real-time troubleshooting guidance for specific service issues.

This multi-level approach dramatically improved performance by transforming decision-making across all organizational levels, not just at the top.

The Role of Decision Processes

Organizations often invest heavily in data capabilities without giving equivalent attention to decision processes—the structured methods through which intelligence translates into action. This imbalance results in sophisticated analytics that fail to create substantial practical impact.

Transformative approaches integrate intelligence with redesigned decision processes that specifically incorporate data-driven insights. Key elements include:

1. **Decision routines:** Regular forums where intelligence informs specific choices

2. **Deliberation frameworks:** Structured approaches to interpreting and applying insights

3. **Divergent thinking protocols:** Methods to explore multiple options before converging on a solution

4. **Integration mechanisms:** Techniques for merging data with experience and judgment

5. **Follow-through systems:** Processes that connect decisions to implementation and continuous learning

A global insurance company transformed its underwriting process by redesigning decision frameworks alongside new analytics capabilities. They established weekly portfolio review sessions where teams systematically examined emerging risk patterns, created structured protocols for integrating algorithmic recommendations with underwriter expertise, and implemented follow-up routines that tracked outcome patterns, allowing them to refine both algorithms and guidelines.

This integrated approach not just better data but better decision processes—enabled the technology to enhance underwriting performance far beyond what either improved analytics or process redesign alone could have achieved.

Intelligence Capability Building

Organizations often underinvest in the human capabilities required to leverage advanced analytics, if simply providing access to data and visualization tools will suffice. This capability gap severely limits transformation potential.

Effective approaches balance technological implementation with deliberate development of what I call the "intelligence skill stack", the capabilities required at different organizational levels:

1. **Executive Intelligence Skills:**
   - Formulating strategic questions
   - Recognizing patterns across domains
   - Balancing data with judgment
   - Cultivating data-driven cultures

2. **Management Intelligence Skills:**
   - Translating insights into operational implications
   - Designing decision experiments
   - Facilitating data-informed collaboration
   - Building team analytical capabilities

3. **Professional Intelligence Skills:**
   - Interpreting data critically
   - Understanding relevant statistics
   - Communicating insights effectively
   - Connecting analytics to domain expertise

A healthcare organization implementing predictive analytics for patient outcomes systematically built these capabilities alongside their technical infrastructure. They designed tiered learning experiences tailored to different roles—from executive interpretation sessions focused on strategic application to frontline clinician workshops on integrating predictive insights with clinical judgment.

This capability-building approach enabled the organization to extract substantially more value from the same technical infrastructure compared to peer organizations that primarily focused on technological implementation.

The Insight Translation Challenge

Even with good data and capable people, organizations often struggle with insight translation, moving from analytical findings to practical application. This translation gap prevents intelligence from becoming transformative action.

Effective approaches address this challenge through what I call insight activation systems, bridges that connect analytical work to operational reality. Key components include:

1. **Insight distillation:** Processes that extract actionable insights from complex analysis
2. **Contextual framing:** Methods for connecting insights to specific business scenarios
3. **Option development:** Techniques for translating findings into practical strategies
4. **Implementation pathways:** Clear connections between insights and execution mechanisms
5. **Learning loops:** Mechanisms for tracking what happens when insights are applied

A retail organization established a dedicated "insight activation team" that partnered with data scientists to translate analytical findings into specific merchandising and marketing strategies. This team, which combined analytical understanding with deep operational knowledge, created implementation playbooks that helped store managers apply customer behaviour insights in practical, actionable ways that data scientists alone could not have developed.

This approach dramatically improved the ROI on their analytics investment by ensuring insights consistently led to value-creating decisions.

The Intelligent Organization Model

Beyond specific tools and teams, transformative approaches develop what I call "the intelligent organization model", a framework that enables data-informed adaptation at scale:

1. **Inquiry culture:** Norms that encourage questioning, exploration, and evidence-based thinking
2. **Decision transparency:** Visibility into how and why important organizational choices are made
3. **Learning discipline:** Structured approaches for extracting insights from experience and outcomes
4. **Knowledge networks:** Systems that connect insights across teams and functions
5. **Intelligence democratization:** Broad access to data and analytical capabilities

A global manufacturer systematically built these characteristics through deliberate interventions: they established "inquiry rituals" where leaders publicly examined their assumptions, created decision records explicitly linking choices to supporting evidence, implemented after-action review processes for all major initiatives, built communities of practice that shared analytical insights across locations, and developed tiered data literacy programs to enable employees at all levels to engage with relevant information.

This holistic approach created an organizational framework where isolated intelligence initiatives could connect, amplify, and compound, generating transformation beyond what any single analytics project could achieve alone.

Ethical Intelligence

As organizations develop more sophisticated data capabilities, they increasingly face complex ethical questions regarding data collection, analysis, and application. Addressing these questions is not just a compliance requirement but a critical enabler of transformation—building the trust necessary for intelligent action.

Transformative approaches develop "ethical intelligence frameworks"—structured guidelines for navigating these complexities. Key elements include:

1. **Data ethics principles:** Clear values guiding decisions about data collection and usage

2. **Algorithmic impact assessment:** Processes to evaluate potential risks of automated systems

3. **Transparency standards:** Guidelines for when and how to explain data-driven decisions

4. **Augmentation boundaries:** Clarity on where human judgment remains essential

5. **Stakeholder voice mechanisms:** Methods for incorporating diverse perspectives in intelligence design

A healthcare organization developing predictive analytics for employee turnover established a comprehensive ethical framework that defined strict boundaries on what data could be collected, mandated explainability for algorithmic recommendations, ensured human oversight in personnel decisions, and created an ethics review board with representatives from diverse employee groups to evaluate all new analytics approaches.

This ethical approach actually accelerated their transformation by proactively addressing misuse concerns, which had stalled previous analytics initiatives, and establishing trust for broader adoption.

Intelligence for HR Transformation

HR functions have unique opportunities and challenges in building intelligence capabilities. Effective approaches focus on several key dimensions:

1. The Multiple Value Lenses

HR intelligence must simultaneously serve multiple stakeholder perspectives:

- **Operational lens:** Enhancing efficiency and effectiveness of HR processes
- **Talent lens:** Improving people-related decision-making and development
- **Strategic lens:** Connecting workforce capabilities to business outcomes
- **Experience lens:** Understanding and enhancing employee experience

A global technology company built its HR intelligence architecture around these four lenses, ensuring their capabilities served diverse needs rather than optimizing for a single perspective.

2. The Insight-to-Action Bridge

HR teams often produce sophisticated analytics that fail to translate into actionable talent decisions for managers. Transformative approaches deliberately bridge HR insights and operational execution by:

- **Decision embedding:** Integrating intelligence into workflow moments where decisions **occur**
- **Manager enablement:** Equipping line leaders to apply workforce insights effectively
- **Insight translation:** Converting complex analysis into practical, tailored implications
- **Collaborative sensemaking:** Creating cross-functional forums where HR and operations jointly interpret workforce data

A manufacturing organization transformed its retention strategy by moving beyond predictive turnover reports. Instead, they embedded guided decision support into manager tools, created "retention dialogue guides" to help leaders apply insights in conversations, and established monthly talent forums where HR and operations collaboratively analyzed engagement patterns and designed interventions.

3. The Ethics Imperative

HR must lead in establishing ethical intelligence practices to maintain trust while supporting transformation. This leadership includes:

- Defining clear principles for people data collection and use
- Ensuring transparency in algorithmic talent decisions
- Maintaining human oversight in consequential workforce matters
- Engaging diverse voices in workforce analytics design
- Monitoring unintended consequences in automated systems

A financial services organization developed "People Analytics Ethics Principles" that explicitly defined boundaries around employee data usage, required ethics reviews for all algorithmic HR tools, and formed a diverse ethics council to evaluate analytics applications before implementation.

This ethical commitment positioned HR as a trusted advisor in the organization's broader intelligence transformation, rather than just another function implementing analytics.

Intelligence as a Transformation Accelerator

When developed effectively, the intelligence layer serves as a powerful accelerator for broader organizational transformation. This acceleration happens through:

1. **Reality feedback:** Providing accurate insight into transformation progress and gaps
2. **Pattern recognition:** Identifying where change is succeeding or struggling
3. **Resource optimization:** Enabling targeted allocation of transformation support

4. **Hypothesis testing:** Validating which aspects of the transformation are working

5. **Narrative enrichment:** Creating compelling evidence-backed transformation stories

A retail organization implementing customer experience technologies used intelligence to significantly accelerate adoption. By analyzing technology usage patterns alongside customer satisfaction metrics, they identified high-impact practices, redirected coaching resources to underperforming locations, tested different training methods, and developed evidence-based narratives that motivated engagement.

This intelligence-accelerated strategy enabled them to achieve in nine months what peer organizations typically required two years to accomplish.

From Insights to Transformation

Ultimately, the intelligence layer transforms organizations not through raw data itself but through how it shapes conversations, influences decisions, and drives actions. This transformation occurs when organizations develop what I call "insight-driven operating models" that embed intelligence into their core ways of working.

Key characteristics of these operating models include:

1. **Hypothesis-driven planning:** Treating strategic and operational decisions as testable hypotheses.

2. **Evidence-based deliberation:** Explicitly incorporating data into critical discussions and choices.

3. **Outcome-oriented metrics:** Measuring what truly matters rather than what's easiest to measure.

4. **Learning-focused reviews:** Using performance data for improvement rather than for judgment.

5. **Insight-embedded processes:** Building intelligence directly into fundamental work processes.

A professional services firm exemplified this approach in their client service transformation. They restructured their entire operating model around intelligence-driven practices: their strategy process began with explicit hypotheses about market opportunities, account teams integrated structured evidence into planning discussions, client satisfaction measurement shifted from activity metrics to relationship outcomes, performance reviews prioritized learning from client interaction patterns, and their core delivery methodology incorporated continuous insight gathering and application.

This holistic redesign of how work functioned—not just better data or tools—enabled technology to genuinely transform their business rather than merely digitalizing existing approaches.

As we transition to the next chapter on HR's evolving role in transformation, remember this essential insight: True transformation demands not just data infrastructure but a comprehensive intelligence layer that connects information to action through human capability, well-designed processes, and organizational support systems. When we invest as intentionally in this intelligence layer as we do in technological platforms, we lay the foundation for data to achieve its full transformative potential.

9

HR as Transformation Agents: Your New Role

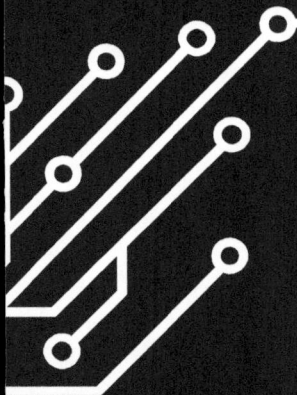

CHAPTER 9

HR as Transformation Agents: Your New Role

When the CEO declared their "digital HR transformation," the HR team immediately focused on selecting and implementing a new human resource information system. Eighteen months later, and after several million dollars, they had successfully moved their existing processes online, enhancing efficiency and expanding data access. Yet, the function's strategic impact and the employee experience remained largely unchanged.

This scenario—mirrored in countless HR functions—illustrates a fundamental misconception of HR transformation. True transformation isn't about merely digitizing existing HR processes; it's about fundamentally redefining HR's role in a technology-enabled organization.

In this chapter, we'll explore how HR can evolve from service providers and program administrators to transformation agents who drive broader organizational change while simultaneously revolutionizing their own function.

The Evolution of HR's Transformation Role

HR's engagement with transformation has evolved through three distinct phases, each reflecting broader shifts in how organizations approach technology-enabled change:

Phase 1: HR as Technology Implementers

In this initial phase, HR focused primarily on implementing HR technology systems to improve administrative efficiency. Success was measured by technical deployment metrics, cost reduction, and process standardization.

The key shortcomings of this approach were:

- Technology automated existing HR processes rather than reinventing them.
- Implementation prioritized HR efficiency over business impact.
- Employee experience was secondary to administrative cost savings.
- Technology remained confined within the HR function.

Phase 2: HR as Service Transformers

As technology advanced, HR shifted focus from pure implementation to service transformation—leveraging technology to redefine service delivery models and enhance stakeholder experiences. Success metrics expanded to include adoption rates, employee satisfaction, and service-level improvements.

Despite its progress, this approach still had significant limitations:

- Transformation remained centred on HR's own services.
- Technology improved service delivery but contributed minimally to business transformation.
- HR's strategic impact remained constrained despite increased efficiency.
- Efforts focused on HR's internal processes rather than enhancing organizational capabilities.

Phase 3: HR as Transformation Agents

In the most advanced phase—still emerging in many organizations—HR transcends service transformation to become a key driver of enterprise-wide change. In this role, HR leverages its human expertise and cross-functional perspective to shape how technology enables new ways of working, leading, and creating value.

This evolution broadens HR's transformation scope:

- From HR processes to enterprise work patterns and capability-building.
- From service efficiency to human-centred value creation.
- From program implementation to ecosystem orchestration.
- From supporting change to architecting transformation.

A global manufacturer exemplified this shift when implementing a new workforce management system. Rather than focusing solely on scheduling efficiency, their HR team recognized a broader transformation opportunity. They facilitated cross-functional dialogues about how flexible staffing could enhance customer responsiveness, designed reflection processes to capture emerging insights about skill deployment, and established learning communities where managers shared best practices for balancing algorithmic recommendations with human judgment.

This expanded approach transformed not just scheduling administration but fundamental aspects of how the organization allocated talent to drive business value—impacting the enterprise far beyond what a narrower implementation focus could have achieved.

The Five Roles of HR as Transformation Agents

As HR evolves into transformation agents, they take on five interconnected roles that transcend traditional functional boundaries:

1. Human Experience Architects

HR possesses deep expertise in human psychology, motivation, and social dynamics, which is crucial for ensuring that technology implementations enhance rather than diminish the human experience of work.

In this role, HR:

- **Evaluates** how technological changes impact the psychological and social aspects of work
- **Develops** implementation strategies that support human adaptation
- **Creates** meaning-making processes that align change with organizational purpose
- **Ensures** that technology fosters human connection instead of eroding it

A healthcare organization's HR team embodied this role when rolling out a new patient management system. They established "experience labs" where clinicians explored the technology's effects on patient interactions, designed implementation strategies that protected key relationship moments while automating administrative tasks, and facilitated purpose-driven dialogues to frame the system as a tool for care quality rather than just efficiency.

2. Capability Catalysts

HR's expertise in learning, development, and talent management positions them to build the capabilities essential for technology-enabled transformation.

In this role, HR:

- **Identifies** the capability requirements for transformation initiatives
- **Develops** integrated strategies for building critical skills

- **Fosters** social learning and knowledge-sharing environments
- **Promotes** "learning agility" as an essential organizational competency

A retail company's HR team redefined capability building during a customer experience technology rollout. Instead of simply training employees on the system, they created store-based learning communities, enhanced managers' ability to coach technology application, hosted peer showcases for innovative approaches and implemented real-time feedback loops to accelerate adaptation.

3. Culture Shapers

HR, as stewards of organizational culture, plays a key role in shaping the cultural "operating system" that determines how effectively technology drives transformation.

In this role, HR:

- **Defines** how cultural characteristics enable or hinder transformation
- **Aligns** talent practices with a transformation-supportive culture
- **Uses** storytelling and symbolic actions to drive cultural evolution
- **Guides** leaders in modelling the behaviours essential for change

A financial services firm's HR team led a deliberate culture shift alongside their digital transformation. They revamped recognition programs to celebrate collaborative innovation, introduced storytelling channels that highlighted cross-functional problem-solving, facilitated leadership conversations to shift from control to enablement, and integrated cultural indicators into transformation metrics.

4. Ecosystem Orchestrators

HR's enterprise-wide perspective and cross-functional influence enable them to orchestrate the complex ecosystem of stakeholders, capabilities, and initiatives needed for successful transformation.

In this role, HR:

- **Maps** interdependencies across transformation initiatives
- **Connects** previously siloed efforts into a cohesive strategy
- **Facilitates** cross-functional collaboration and learning
- **Ensures** integration mechanisms create a seamless change experience

A global manufacturing firm's HR team exemplified this role in their Industry 4.0 transformation. They formed cross-functional design teams to align technology adoption with work process redesign, established transformation learning forums, developed coordination mechanisms to prevent initiative collisions, and designed integrated metrics tracking progress across technical, human, and business dimensions.

5. Decision Scientists

HR's access to workforce data uniquely positions them to apply evidence-based approaches to transformation, shifting from intuition to insight-driven strategies.

In this role, HR:

- **Collects** and analyses data on the human dimensions of transformation
- **Designs** experiments to refine implementation strategies

- **Develops** measurement systems to track transformation progress
- **Builds** the organization's capability for evidence-based decision-making

A technology company's HR team exemplified this role during their remote work transition. They conducted natural experiments on team structures, created advanced metrics tracking both performance and experience, deployed pulse surveys to surface emerging challenges, and built predictive models to identify teams likely to struggle with virtual collaboration. This scientific approach led to more precise interventions and strengthened the organization's capacity for data-driven transformation.

The Transformation Agent Mindset

Beyond specific roles, HR professionals must adopt a mindset shift to become true transformation agents. This shift includes:

1. From Function to Enterprise

Traditional HR prioritizes its own processes; transformation agents align HR with broader organizational goals.

A pharmaceutical company's HR leadership embodied this shift during their clinical transformation. Rather than focusing on HR system implementation, they immersed themselves in clinical work changes, assessed how technology reshaped coordination, and optimized talent practices to support new research models.

2. From Programs to Platforms

Traditional HR delivers one-off programs; transformation agents build platforms for continuous learning and innovation.

A professional services firm's HR team abandoned a rigid "digital leadership program", instead creating a leadership acceleration platform with peer learning communities, real-time feedback systems, curated learning tools, and project-based development.

3. From Best Practices to Emergent Learning

Traditional HR relies on benchmarking; transformation agents experiment, learn, and iterate in real time.

A retail company's HR team replaced rigid best practices with store innovation labs where teams experimented with new approaches, implemented rapid feedback loops, and shared insights faster than formal training rollouts.

4. From Service Provider to Strategic Catalyst

Traditional HR responds to requests; transformation agents proactively shape change strategies.

A manufacturing firm's HR team preemptively facilitated strategic discussions about how connected machines would reshape skills, decision-making, and organizational design, ensuring these considerations were integrated early rather than as afterthoughts.

5. From Prescriptive to Experimental

Traditional HR delivers predefined solutions; transformation agents embrace iterative, experimental implementation.

A healthcare organization's HR team tested multiple care team structures during their telehealth transformation, capturing real-world insights before rolling out final models, ensuring greater adaptability and effectiveness.

Transforming HR for Transformation Impact

To fulfill these expanded roles effectively, HR must simultaneously transform itself. This parallel transformation involves several key dimensions:

1. Reimagined HR Structure

Traditional HR structures, which are organized around functional specialties such as compensation, learning, and talent acquisition, often struggle to deliver integrated transformation support. More effective structures instead include:

- **Human experience teams** organized around employee journeys or moments that matter
- **Business transformation partners** embedded in major change initiatives
- **Capability acceleration centers** that build critical skills across boundaries
- **People analytics centers of excellence** that provide decision science expertise

A global technology company explicitly redesigned their HR structure to better support transformation. They created specialized journey teams responsible for integrated employee experiences, embedded HR strategists within major transformation initiatives, established a capability accelerator that designed cross-functional learning experiences, and built an insights team that provided analytics expertise across the enterprise.

2. New HR Capabilities

Supporting transformation requires capabilities many HR functions historically haven't fully **embraced:**

- **Systems thinking:** Understanding complex interdependencies and feedback loops

- **Design thinking:** Creating human-centered experiences and solutions
- **Data literacy:** Interpreting and applying analytics to transformation challenges
- **Digital fluency:** Understanding technology implications for work and organizations
- **Change architecture:** Designing conditions for successful adaptation

A financial services organization invested significantly in building these capabilities within their HR team. They created immersive learning experiences where HR professionals practiced systems mapping of transformation initiatives, participated in human-centred design workshops for employee experiences, completed digital literacy programs focused on emerging technologies, and developed practical change architecture skills through guided application.

3. Integrated Technology Strategy

HR's technology approach must evolve beyond automating HR processes toward actively enabling enterprise transformation. This evolution includes:

- **Ecosystem thinking,** which connects HR systems to broader enterprise platforms
- **Experience design,** ensuring intuitive, context-aware interactions
- **Intelligence architecture,** which transforms data into actionable insights
- **Platform approaches,** enabling continuous adaptation and extension

A professional services firm exemplified this evolution when implementing their new HRIS. They designed the system as an enterprise-wide people platform rather than an HR-specific administrative tool, created customized experiences for various user types based on their specific contexts, built sophisticated analytics capabilities that surfaced actionable workforce insights, and established a continuous enhancement approach that enabled ongoing evolution rather than relying on periodic major upgrades.

4. Collaborative Methods

Traditional HR often operates through formal governance and standardized processes. Transformation support requires more collaborative, adaptive approaches:

- **Co-creation methods** that engage diverse stakeholders in solution design
- **Adaptive implementation** that evolves approaches based on feedback
- **Community activation** that energizes informal networks alongside formal structures
- **Boundary-spanning forums** that connect previously isolated initiatives

A healthcare organization's HR team transformed how they approached a major workforce redesign by adopting these collaborative methods. They created cross-role design teams that shaped new care models together, implemented feedback mechanisms that enabled continuous adaptation during rollout, activated informal physician and nurse networks to spread emerging practices, and established integration forums that connected the workforce initiative with parallel technology and process changes.

5. Balanced Metrics

Traditional HR metrics often focus primarily on functional efficiency, cost, and compliance. Transformation agents instead develop more balanced measurement approaches that include:

- **Experience indicators**, tracking the human impact of change
- **Capability metrics**, measuring skill development and application
- **Adaptation markers**, assessing how effectively the organization is evolving
- **Value realization**, linking people dimensions to business outcomes

A retail organization adopted this balanced approach for their customer experience transformation. Alongside traditional implementation metrics, they monitored employee experience during the change, measured the development and application of critical digital service skills, assessed how effectively different stores adjusted their operations, and explicitly connected human factors to customer satisfaction and sales outcomes.

Leading Your HR Transformation Journey

For HR leaders seeking to evolve into transformation agents, the journey requires parallel tracks of evolving internal HR functions while elevating HR's impact on enterprise transformation. Key steps include:

1. Assess Your Current State

Begin by honestly evaluating your HR function's current standing across the five transformation agent roles and internal transformation dimensions. Consider:

- Where do we currently focus most of our transformation energy?
- Which of the five roles are we playing effectively today?
- What capability gaps limit our transformation impact?
- How aligned is our structure and technology in supporting enterprise transformation?

- What mindset shifts do we need to make individually and collectively?

This assessment establishes a clear starting point for evolution.

2. Choose Strategic Entry Points

Rather than attempting a comprehensive transformation immediately, identify specific initiatives where the transformation agent approach can be demonstrated while building capability:

- **Major technology implementations**, where human factors will determine success
- **Work redesign efforts**, requiring integrated perspectives
- **Cultural evolution**, driven by strategic shifts
- **Capability gaps**, limiting organizational adaptation

A manufacturing company's HR team strategically focused its initial transformation agent efforts on a factory automation initiative where they demonstrated the value of integrating human experience design, capability building, and cultural considerations alongside technological implementation.

3. Build Coalitions Beyond HR

Transformation agency requires strong partnerships outside HR's traditional boundaries. Deliberately cultivate relationships with:

- **Technology leaders**, understanding the human side of digital transformation
- **Operations experts**, interested in work redesign opportunities
- **Finance partners**, appreciating the value of human capital investment
- **Customers experience leaders**, recognizing employee experience connections

A healthcare organization's HR leadership built a powerful "transformation coalition," including their CIO, patient experience director, clinical innovation leader, and finance director. This cross-functional partnership enabled broader impact than HR could have achieved operating within traditional boundaries.

4. Develop Showcase Examples

To build momentum and demonstrate the transformation agent approach, create vivid examples of this new role in action:

- Document case studies of integrated transformation initiatives
- Measure and communicate multidimensional outcomes
- Capture stories highlighting the human side of technology-enabled change
- Create immersive experiences that allow others to see new approaches firsthand

A financial services firm's HR team developed a compelling showcase of their transformation approach within wealth management. They documented how integrated attention to advisor experience design, capability building, and cultural evolution improved both advisor satisfaction and client outcomes compared to previous technology-first approaches.

5. Transform Your Own Function

While supporting enterprise transformation, simultaneously evolve HR's own operating model:

- **Redesign structure** to align with transformation needs
- **Develop critical capabilities** through experiential learning
- **Implement technology** that enables an expanded role

- **Track transformation impact** with balanced metrics
- **Create opportunities** for HR professionals to practice new approaches

A global retailer undertook a comprehensive transformation of HR, reorganizing around employee journeys, establishing a digital academy, implementing a sophisticated people analytics platform, and developing scorecards tracking both HR effectiveness and transformation impact.

HR's Unique Opportunity

Perhaps most importantly, HR leaders must recognize the unique opportunity that technology-enabled transformation presents for elevating their strategic contribution. While many perceive technology as a threat to HR's relevance, the reality is precisely the opposite: as organizations increasingly recognize that technology alone cannot deliver transformation, HR's expertise in the human dimensions of change becomes more crucial, not less.

This opportunity requires abandoning defensive postures focused on preserving HR's traditional territory and instead embracing an expansive vision of how HR expertise can shape technology-enabled transformation throughout the enterprise.

A global professional services firm's CHRO exemplified this expansive vision when their organization undertook a comprehensive digital transformation. Rather than focusing narrowly on HR process digitization, she positioned her function as architects of the human experience of transformation. She elevated discussions from technology implementation to human adaptation, from efficiency metrics to experience design, and from project management to cultural evolution.

This strategic elevation transformed perceptions of HR from administrative support to transformation leadership, not by minimizing technology's importance but by illuminating how human factors determine whether technology delivers its transformative potential.

As we turn to the next chapter on the Transformation Canvas, remember this essential truth: HR's opportunity in the digital era isn't to become more technological but to become more essential, applying uniquely human expertise to ensure that technology serves people and purpose rather than the reverse. By evolving from HR digitizers to transformation agents, HR leaders position themselves and their functions as indispensable architects of organizational adaptation in an increasingly technological world.

10

The Transformation Canvas:
A Holistic Framework

CHAPTER 10

The Transformation Canvas: A Holistic Framework

Throughout this book, we've explored multiple dimensions of true transformation beyond technology implementation. We've examined how human psychology, organizational culture, leadership approaches, process design, capability building, change architecture, intelligence systems, and HR's evolving role all contribute to whether technology delivers transformative value.

The challenge for leaders is integrating these dimensions into a coherent approach rather than addressing them as separate workstreams. This integration challenge explains why many organizations achieve excellence in individual aspects of transformation but still fail to realize the full potential of their technology investments.

In this chapter, we'll explore the Transformation Canvas—a holistic framework that helps leaders design and orchestrate true transformation across all essential dimensions. This canvas serves as both a diagnostic tool for identifying gaps in current approaches and a design framework for creating more integrated transformation strategies.

The Need for Integration

Before introducing the canvas, let's examine why integration is so crucial for transformation success:

1. **Interdependency Reality:** Each transformation dimension significantly affects others. Leadership approaches influence cultural evolution; process

designs impact capability requirements; human experiences shape data utilization patterns. When addressed separately, these interdependencies create friction rather than synergy.

2. **Experience Coherence:** From the perspective of those experiencing change, transformation appears as a single integrated journey, not separate workstreams. Fragmented approaches create confusing, sometimes contradictory experiences that increase resistance and reduce engagement.

3. **Resource Optimization:** Siloed transformation efforts frequently duplicate efforts, create competing priorities, and sub-optimize resource allocation. Integrated approaches enable strategic focus on the highest-impact interventions across dimensions.

4. **Accelerated Learning:** Integrated approaches create faster feedback loops and cross-dimensional insights that accelerate adaptation. Siloed efforts miss crucial patterns that emerge at intersection points.

A global manufacturer's digital transformation exemplified this integration challenge. Their technology implementation and process redesign teams created sophisticated production optimization systems, but their leadership development, culture building, and capability efforts operated separately with different timelines and approaches. The result was technically excellent systems that delivered minimal performance improvement because the human and organizational dimensions weren't ready to leverage the technology effectively.

The Transformation Canvas Structure

The Transformation Canvas provides a visual framework for designing integrated transformation across nine essential dimensions organized in three layers:

Foundation Layer

1. **Purpose & Direction:** Why transformation matters and where it's headed

2. **Human Experience:** How transformation affects and is shaped by people

3. **Leadership & Culture:** How transformation is guided and enabled

Activation Layer

4. **Process & Work Design:** How value-creating work is reimagined

5. **Capability Building:** How needed skills and competencies are developed

6. **Change Architecture:** How adaptation is structured and supported

Intelligence Layer

7. **Data & Decisions:** How insights inform and enhance transformation

8. **Learning Systems:** How the organization captures and applies what it learns

9. **Value Realization:** How transformation creates measurable impact

For each dimension, the canvas prompts leaders to address key questions, design focused interventions, and identify integration points with other dimensions.

Using the Transformation Canvas

The canvas can be used in three primary ways:

1. **Diagnostic Assessment:** Evaluating current transformation approaches to identify gaps, imbalances, and integration opportunities

2. **Design Framework:** Creating comprehensive transformation strategies that address all essential dimensions

3. **Alignment Tool:** Building shared understanding and coordinated action across transformation teams

Let's explore how the canvas works in practice by examining each dimension and its integration points.

Foundation Layer: Purpose & Direction

Key Questions

- Why does this transformation matter to our organization and stakeholders?
- What specific outcomes are we trying to achieve?
- What principles will guide our transformation journey?
- How will we know if we succeed?

Integration Points

- **Human Experience:** How purpose connects to meaningful human motivations
- **Leadership & Culture:** How direction aligns with and shapes cultural values
- **Value Realization:** How purpose translates to measurable outcomes

A healthcare organization used the Purpose & Direction element of the canvas to reframe their electronic health record implementation. Rather than positioning it as a technical upgrade, they articulated a compelling purpose around "creating seamless care journeys for patients" and established principles emphasizing that technology would enhance rather than replace human connections in healthcare. This purpose framing dramatically affected how clinicians engaged with the initiative compared to peer organizations that emphasized technical compliance.

Foundation Layer: Human Experience

Key Questions

- How will transformation affect different stakeholders' experiences?
- What emotions and psychological responses might emerge?
- How might we design changes with human adaptation in mind?
- Where are the moments that matter most in the transformation journey?

Integration Points

- **Process & Work Design:** How work redesign affects day-to-day experience
- **Change Architecture:** How support structures address psychological needs
- **Data & Decisions:** How human factors influence data interpretation and use

A financial services company used the Human Experience element to transform their approach to advisor platform implementation. Rather than focusing solely on functionality, they mapped the emotional journey advisors would experience transitioning from familiar to new systems. This analysis revealed critical "moments of truth" where targeted support would be essential and identified experience opportunities where the technology could enhance rather than disrupt valued client relationships.

Foundation Layer: Leadership & Culture

Key Questions

- What leadership mindsets and behaviors will enable transformation?

- Which cultural characteristics will help or hinder our efforts?

- How will decision rights and power dynamics need to evolve?

- What symbols and stories will reinforce desired changes?

Integration Points

- **Capability Building:** How leadership development supports new approaches

- **Change Architecture:** How cultural factors influence adaptation structures

- **Learning Systems:** How leadership enables or constrains organizational learning

A manufacturing organization used the Leadership & Culture element to address why their past technology implementations had consistently underperformed. They identified that their top-down, command-oriented leadership style and deeply ingrained risk-averse culture were fundamentally incompatible with the decentralized, innovation-driven mindset their new systems were designed to enable. This insight led them to implement targeted leadership development initiatives, symbolic organizational shifts, and structural transformations that aligned cultural evolution with their technology implementation efforts.

Activation Layer: Process & Work Design

Key Questions

- How will fundamental, value-creating work be redesigned, rather than merely digitized?

- Where can technology eliminate, automate, augment, or transform work?

- How will roles, responsibilities, and workflows evolve?
- What new collaboration and coordination mechanisms will be required?

Integration Points

- **Human Experience:** How redesigned work impacts meaningful employee engagement
- **Capability Building:** What new skills redesigned processes require
- **Data & Decisions:** How workflow generate, interpret, and apply intelligence

A retail organization used the Process & Work Design element to move beyond the basic adoption of customer analytics technology. They fundamentally reimagined their entire merchandising approach— transitioning from rigid, season-based planning cycles to an agile, continuously adaptive model. This shift also involved moving from a centralized, top-down decision-making structure to a distributed, data-informed approach and transitioning from standardized product assortments to localized, customer-driven selections. It wasn't just the technology itself that delivered breakthrough results, it was this radical rethinking of processes that drove their performance transformation.

Activation Layer: Capability Building

Key Questions

- What specific new skills and competencies will be required for successful transformation?
- How will we build capabilities at individual, team, and organizational levels?
- What dynamic learning mechanisms will support ongoing skill development?
- How can we accelerate time-to-competence in mission-critical areas?

Integration Points

- **Process & Work Design:** How capability enables new work approaches

- **Change Architecture:** How learning integrates with change support

- **Leadership & Culture:** How capability building influences leadership mindsets and behaviors

A pharmaceutical company leveraged the Capability Building element to transform their approach to clinical trial technology implementation. Rather than focusing narrowly on technical system training, they developed a holistic capability-building strategy that encompassed data literacy, remote collaboration techniques, and adaptive leadership skills—alongside technical proficiency. This comprehensive, multi-layered approach enabled teams to fully leverage the technology's potential, outperforming competitors who had limited their focus to software training alone.

Activation Layer: Change Architecture

Key Questions

- How will we create conditions for effective, sustainable adaptation?

- What structures will support people through the transformation journey?

- How will we balance structured guidance with emergent, co-created solutions?

- What iterative feedback mechanisms will help us refine and adjust our approach in real time?

Integration Points

- **Human Experience:** How change supports addresses emotional and psychological responses

- **Learning Systems:** How adaptation insights feed continuous improvement
- **Leadership & Culture:** How change architecture reinforces leadership philosophies

A telecommunications company used the Change Architecture element to transform the rollout of its digital customer experience initiative. Rather than relying on a rigid, centralized change management function that focused narrowly on communication and training, they established networked change teams with localized decision-making autonomy, peer-driven support communities that fostered collective learning, and real-time feedback loops that enabled constant course correction. This decentralized, responsive approach significantly improved adoption rates and long-term success.

Intelligence Layer: Data & Decisions

Key Questions

- How will data-driven insights inform and enhance transformation efforts?
- What decision processes will harness intelligence to drive better outcomes?
- How will we cultivate organization-wide data literacy and analytical capabilities?
- What ethical frameworks will guide our use of information?

Integration Points

- **Process & Work Design:** How workflows integrate data into decisions
- **Human Experience:** How insights can enhance decision-making without overwhelming users
- **Capability Building:** How analytical skills develop throughout the organization

A healthcare organization used the Data & Decisions element to revolutionize their approach to patient safety technology. They designed sophisticated intelligence systems capable of identifying patterns across incidents, developed AI-powered decision support tools to assist clinicians in applying insights to real-time scenarios, and launched data literacy programs that empowered staff at all levels to interpret and use analytics effectively. Additionally, they implemented strong ethical safeguards to ensure responsible handling of sensitive patient data.

Intelligence Layer: Learning Systems

Key Questions

- How will we capture insights and lessons learned throughout the transformation journey?
- What structures will help translate these insights into actionable improvements?
- How will knowledge flow seamlessly across different teams, departments, and leadership levels?
- How will we balance structured learning with organic, emergent adaptation?

Integration Points

- **Change Architecture:** How real-time learning feeds continuous transformation improvements
- **Leadership & Culture:** How learning systems reflect and reinforce cultural values
- **Data & Decisions:** How experiential learning complements data-driven insights

A financial services firm used the Learning Systems element to significantly enhance the effectiveness of their digital transformation efforts. They established deliberate, structured learning mechanisms, including frontline feedback loops, interdepartmental knowledge-sharing forums, structured reflection sessions, and ongoing peer learning communities. This intentional focus on cross-functional learning accelerated organizational adaptation, in stark contrast to their previous transformation approach, where each team operated in isolation, resulting in fragmented and inefficient progress.

Intelligence Layer: Value Realization

Key Questions

- How will we measure the impact of transformation across multiple dimensions?

- What early indicators will help us adjust course early?

- How will we balance short-term and long-term value creation?

- How will we accurately attribute value to different transformation elements?

Integration Points

- **Purpose & Direction:** How value metrics connect to transformation purpose

- **Data & Decisions:** How measurement informs ongoing adaptation

- **Leadership & Culture:** How value definition reflects organizational priorities

A manufacturing organization used the Value Realization element to fundamentally rethink how they assessed the effectiveness of their Industry 4.0 implementations. Rather than focusing exclusively on efficiency-related KPIs, they developed a comprehensive balanced scorecard that encompassed operational performance, financial outcomes, human experience, and innovation impact. This broader evaluation framework enabled them to identify unexpected yet highly valuable outcomes, leading to more strategic decision-making about resource allocation in future implementation phases.

The Integration Matrix

Beyond addressing individual dimensions, the true power of the canvas emerges when leaders identify and activate the integration points between dimensions. The canvas includes an Integration Matrix that explicitly maps these connections and prompts leaders to design interventions that span multiple dimensions.

For example, the intersection of Human Experience and Process & Work Design might prompt questions like:

- How does our process of redesign affect meaningful aspects of stakeholder experiences?
- Where might technical efficiency and human experience come into tension?
- How can we design processes that optimize both operational and human outcomes?

A retail organization used this intersection analysis approach to radically enhance their customer service technology implementation. They discovered that while their process redesign improved transaction speed and efficiency, it had inadvertently eliminated key customer interaction elements that frontline employees found most fulfilling. By restructuring their approach to preserve these meaningful moments—while still benefiting from efficiency gains—they achieved significant improvements in both customer satisfaction and employee engagement.

Using the Canvas in Practice

The canvas can be applied at multiple stages of transformation, with each phase requiring a slightly different focus and application strategy:

1. Initial Assessment

At the early stages, use the canvas to evaluate current transformation readiness and identify critical gaps across dimensions. Key activities include:

- Rating current capability in each dimension
- Identifying the biggest gaps requiring attention
- Mapping interdependencies between dimensions
- Prioritizing initial focus areas based on assessment

2. Transformation Design

Use the canvas to create a comprehensive transformation strategy that addresses all dimensions. Key activities include:

- Evaluating and rating the organization's current capability across each dimension
- Pinpointing the most significant gaps requiring urgent attention
- Mapping interdependencies between transformation dimensions
- Prioritizing initial focus areas based on the assessment findings

3. Implementation Adaptation

During implementation, use the canvas to monitor progress and adjust approach. Key activities include:

- Assessing which dimensions are progressing effectively and which are lagging
- Identifying emerging interdependencies that require additional focus
- Adjusting resource allocation and strategic priorities based on evolving needs
- Capturing and leveraging cross-dimensional learning insights

A global professional services firm used canvas throughout their digital workplace transformation. Initial assessment revealed strong approaches to technology and process dimensions but significant gaps in cultural evolution and human experience design. Their transformation design integrated these previously neglected dimensions, and their implementation approach included regular canvas-based reviews that helped them identify and address emerging challenges at dimension intersection points.

Canvas-Based Transformation Governance

Beyond one-time application, many organizations implement ongoing transformation governance based on the canvas framework. This typically includes:

1. **Dimension Stewards:** Leaders responsible for ensuring excellence within specific canvas dimensions

2. **Integration Teams:** Cross-functional groups focused on key dimension intersections

3. **Canvas Reviews:** Regular sessions evaluating progress across all dimensions

4. **Adjustment Mechanisms:** Processes for reallocating resources based on canvas insights

A healthcare organization implemented this governance approach for their patient care transformation. They assigned executive sponsors to each canvas dimension, created teams for critical intersections like technology-culture and process-capability, conducted quarterly canvas reviews that assessed progress holistically, and maintained flexible transformation funding that could be reallocated based on canvas insights.

This integrated governance approach enabled them to maintain balance across dimensions and address emerging challenges before they became critical barriers.

The Canvas as Transformation Language

Perhaps most importantly, the canvas provides organizations with a shared language and mental model for transformation. This common framework enables more sophisticated discussions about transformation challenges and opportunities.

Rather than fragmented conversations where technology teams discuss implementation timelines, HR discusses change management, and operations discusses process efficiency, the canvas enables integrated dialogue about how these dimensions interact and how the organization can orchestrate them for maximum impact.

A manufacturing organization used the canvas framework to transform how their executive team discussed and led their digital transformation. Instead of separate updates on technology deployment, process redesign, and organizational readiness, their leadership discussions were organized around the canvas dimensions with particular focus on integration points and emerging patterns. This integrated perspective led to several crucial adjustments, including slowing technical implementation to allow cultural and capability dimensions to catch up—that ultimately delivered much stronger results than their previous technology-first approach.

From Framework to Action

While the canvas serves as a powerful conceptual framework, its true value lies in its ability to drive practical, actionable decisions. Organizations that leverage it most effectively use it to inform key strategic choices, including:

- **Resource Allocation:** Redirecting investments toward dimensions with the greatest gaps
- **Timeline Adjustment:** Restructuring implementation sequencing to align interdependent dimensions more effectively

- **Scope Refinement:** Expanding or narrowing transformation elements based on dimension-level analysis

- **Team Composition:** Integrating additional expertise in areas requiring stronger focus

- **Success Redefinition:** Evolving performance metrics to reflect a more holistic, multidimensional view of transformation success

A financial services organization applied these canvas-driven insights to fundamentally reshape their digital banking transformation strategy. Their initial analysis revealed a disproportionate emphasis on technology and process improvements, while critical aspects such as customer experience design and cultural adaptation were significantly underfunded. In response, they reallocated resources, incorporated experience design specialists into their core transformation team, restructured their implementation timeline to allow cultural evolution to precede technical deployment in key areas, and expanded their success metrics beyond technical adoption to include user experience and behavioural engagement indicators.

The Personal Canvas

Beyond organizational application, the canvas can serve as a powerful tool for individual transformation leaders to assess and develop their own capabilities. By evaluating their personal expertise and comfort across canvas dimensions, leaders can identify growth areas that will make them more effective transformation guides.

A retail CHRO used this personal application to develop her capabilities as a transformation leader. Analysis revealed she was highly skilled in the human dimensions of the canvas but less comfortable with data and technological elements. She created a personal development plan that included technology immersion experiences, analytics mentoring, and participation in digital strategy forums. This balanced capability development significantly increased her effectiveness in guiding her organization's transformation efforts.

As we move to the next chapter on case studies, remember this essential insight: True transformation requires attention to multiple dimensions and, most importantly, to how these dimensions interact and reinforce each other. The Transformation Canvas provides a structured framework for this integrated approach, helping leaders move beyond technology-centric transformation to orchestrate the full spectrum of changes required for technology to deliver its transformative potential.

11

Case Studies:
Real Transformation in Action

CHAPTER 11

Case Studies: Real Transformation in Action

The principles and frameworks discussed throughout this book aren't theoretical constructions, but practical approaches proven in real organizations facing complex transformation challenges. In this chapter, we'll explore four detailed case studies that illustrate what true transformation looks like when organizations move beyond technology-centric approaches to more holistic models.

Each case study examines how organizations address multiple transformation dimensions, overcame significant challenges, and achieved outcomes that technology alone could not have delivered. While the organizations have been anonymized, the stories and lessons are real, drawn from my direct experience advising these transformations.

Case Study 1:
Beyond Digital HR at Global Financial Services

Organization Profile: A global financial services firm with 75,000 employees across 40 countries, operating in investment banking, wealth management, and retail banking.

Transformation Context: The organization had invested $40 million in a new HR technology platform but was struggling to realize significant value beyond administrative efficiency. Employee experience remained fragmented, HR business partners were overwhelmed with transactional work despite self-service capabilities, and talent decisions showed little improvement despite enhanced analytics.

Technology-First Approach and Its Limitations

The organization had approached transformation with a classic technology-first mindset. They had:

- Implemented a leading cloud-based HR platform with extensive self-service and analytics
- Digitized existing HR processes with minimal redesign of underlying work patterns
- Trained employees and managers on system usage without redefining roles or relationships
- Created an HR shared service center focused on transaction processing

While this approach delivered administrative efficiency (reducing HR operating costs by 15%), it failed to transform the employee experience, strategic HR capabilities, or talent outcomes. HR remained perceived as a primarily administrative function despite its significant technology investment.

The Pivot to True Transformation

Recognizing these limitations, the CHRO initiated a fundamental reset of their approach. Key shifts included:

1 Purpose Reframing

Rather than positioning transformation as "HR technology modernization," they reframed it as "creating frictionless employee experiences and insight-driven talent decisions." This purpose shift significantly altered how stakeholders engaged with the initiative.

2 Experience-Centered Redesign

Instead of organizing around HR processes, they mapped critical employee journeys (joining, developing, changing roles, existing) and redesigned each from the employee perspective. This mapping revealed that employees experienced up to 18 separate HR transactions during major life events, despite having a unified technology platform.

The team redesigned these experiences from scratch—creating integrated journeys with simplified processes, proactive guidance, and meaningful human touchpoints where they mattered most. Technology became an enabler of these experiences rather than the starting point for design.

3 HR Role Reimagination

The organization fundamentally reconceived HR roles beyond traditional centers of excellence and business partner models. New roles included:

- **Experience Designers:** HR professionals skilled in journey mapping and human-centered design
- **People Analytics Consultants:** Specialists who helped leaders apply data to talent decisions
- **Talent Strategists:** Forward-looking advisors focused on workforce planning and capability building
- **Culture Navigators:** Experts in shaping organizational culture and change

This role evolution shifted HR from process administrators to capability architects who enabled transformation throughout the business.

4 Intelligence Integration

Rather than creating standalone HR analytics, they integrated people data with business information to create contextual intelligence that enhanced business decisions. Key innovations included:

- Embedding team composition analytics directly into project planning tools
- Integrating performance patterns with customer satisfaction data to identify experienced drivers
- Creating predictive models that connect leadership behaviors to business outcomes

This integrated approach transformed HR data from interesting information to essential business intelligence.

Cultural Activation

Recognizing that technology alone wouldn't change talent decisions, they launched a deliberate cultural initiative focused on evidence-based leadership. This included:

- Leadership immersion experiences in data-informed decision making
- Peer forums where leaders shared how insights had influenced their approaches

- Recognition for leaders who demonstrated data-informed talent practices
- Adjustments to succession and promotion processes to value analytical capability

This cultural work transformed how leaders thought about and used the available technology and information.

Outcomes and Insights

The reimagined approach delivered significantly stronger outcomes than the technology-centric model:

- Employee satisfaction increased from 68% to 88%
- HR business partners reported spending 60% of their time on strategic work (up from 30%)
- Internal talent mobility increased by 42%
- Leaders reported using people data in 78% of significant talent decisions (up from 23%)
- HR influence in business strategy discussions increased substantially based on independent assessment

Key insights from this transformation included:

1. **Technology enables but experiences transform:** The same technology platform delivered dramatically different value when reimagined from an experience perspective.

2. **Role evolution precedes impact evolution:** HR's strategic impact changed only when HR roles fundamentally shifted beyond traditional models.

3. **Intelligence requires integration:** People data created most value when connected to business contexts and decisions rather than treated as a separate analytical domain.

4. **Cultural capability determines technology value:** The same data and tools delivered different value based on leaders' comfort and capability with data-informed decision making.

5. **Purpose drives engagement:** Reframing transformation in human terms rather than technological ones significantly affected how stakeholders engaged and contributed.

As the CHRO reflected: "We initially thought we were transforming our HR technology. We eventually realized we were transforming our entire approach to enabling human capability in the organization. Technology was necessary but represented perhaps 30% of the actual transformation work."

Case Study 2: Manufacturing Work Reimagined

Organization Profile: A global discrete manufacturer with 28,000 employees across 12 major production facilities, producing precision components for automotive and aerospace applications.

Transformation Context: The organization had invested heavily in Industry 4.0 technologies including IoT sensors, advanced analytics, and automated material handling systems. While technical implementation had been successful, operational performance improvements were minimal, worker satisfaction had declined, and the promised transformation of production operations hadn't materialized.

Technology-First Approach and Its Limitations

The organization had implemented technology with a predominantly technical mindset:

- Engineering led implementation with minimal input from production teams
- Technology focused primarily on monitoring and control rather than enabling frontline decision making
- Training concentrated on system operation rather than new work approaches
- Performance metrics remained focused on traditional efficiency measures

This approach created what workers called "digital micromanagement", technology that monitored their work more closely without enhancing their capability or autonomy. Supervisors felt caught between system directives and practical realities, while plant managers struggled to realize the strategic benefits they had been promised.

The Pivot to True Transformation

Eighteen months into implementation, leadership recognized the need for a fundamental reset. Key shifts included:

1 Work Redesign Through Co-Creation

Rather than imposing technology-driven work patterns, they launched a co-creation process where production teams and engineers jointly redesigned work. This process identified:

- Decision points where operator expertise could enhance algorithmic recommendations
- Information needs that would enable better frontline decision making
- Opportunities to automate routine tasks while expanding problem-solving responsibility
- New coordination patterns between previously siloed roles
- This collaborative redesign transformed technology from a control mechanism to an enablement platform that augmented rather than replaced human judgment.

2 Capability Ecosystem Development

The organization built a comprehensive capability ecosystem that developed technical, analytical, and adaptive skills at all levels:

- Production operators received both technical training and analytical problem-solving development
- Supervisors learned facilitative leadership approaches that leveraged team intelligence
- Engineers developed collaborative design skills to work effectively with production experts

- Plant leaders-built capabilities in managing socio-technical systems and continuous adaptation

This ecosystem approach built the human capabilities required to leverage technological capabilities effectively.

3 Cultural and Leadership Evolution

Recognizing that their command-oriented culture was fundamentally misaligned with the collaborative potential of their new technologies, they undertook deliberate cultural evolution:

- Shifted performance reviews from compliance assessment to learning and development
- Redefined supervisor roles from directive oversight to team enablement
- Created psychological safety through celebration of problem identification
- Adjusted metrics to balance efficiency with learning and innovation

These cultural shifts transformed how technology was perceived and applied throughout production operations.

4 Decision Architecture Redesign

The organization completely redesigned their production decision architecture to leverage both human and machine intelligence:

- Algorithms made routine decisions within defined parameters
- Operators could override algorithmic recommendations with documented rationale
- Production teams made collaborative decisions about process improvements
- Cross-functional forums addressed boundary-spanning operational challenges
- Leadership focused on strategic decisions and enabling conditions

This multilevel decision architecture leveraged technology appropriately at each level while maintaining human judgment where it added greatest value.

5 Learning Integration

Rather than treating production as a stable system to optimize, they redesigned operations as a continuous learning system:

- Created regular reflection sessions where teams reviewed both data and experience
- Established cross-plant communities of practice that shared emerging insights
- Implemented rapid experimentation protocols for testing process improvements
- Developed systematic methods for scaling successful innovations across facilities

This learning orientation transformed operations from a target-focused execution system to an adaptive organism that continuously evolved its capabilities.

Outcomes and Insights

The reimagined approach delivered dramatically different results than the technology-centric implementation:

- Productivity improved 23% compared to 4% under the technology-first approach
- Quality metrics showed 36% defect reduction across product lines
- Employee engagement in production roles increased from 61% to 83%
- Innovation metrics showed a 170% increase in implemented improvement ideas
- Unplanned downtime decreased by 47% through predictive maintenance integration

Key insights from this transformation included:

1. **Human-technology synergy over substitution:** The greatest value came from designing technology to enhance human capabilities rather than replace them.

2. **Co-creation dramatically improves outcomes:** Having those who do the work help design how technology integrates into that work led to significantly better operational solutions.

3. **Culture determines technology utilization:** The same technologies created either command-control reinforcement or distributed intelligence depending on the cultural context.

4. **Learning capability equals performance capability:** Facilities that developed stronger learning systems consistently outperformed those focused primarily on execution excellence.

5. **Decision architecture determines value realization:** Technology created most value when organizations thoughtfully designed who would make which decisions with technological support.

The VP of Operations reflected: "We initially saw this as a technological implementation that would transform our operations. We eventually understood it was a human and organizational transformation enabled by technology. That fundamental shift in perspective made all the difference in our results."

Case Study 3: Healthcare Experience Transformation

Organization Profile: A regional healthcare system comprising 7 hospitals, 120+ outpatient facilities, and 18,000 employees serving a diverse patient population across urban and rural settings.

Transformation Context: The organization had implemented a comprehensive electronic health record and patient engagement platform but was struggling with provider burnout, fragmented patient experiences, and minimal improvement in care coordination despite significant technology investment.

Technology-First Approach and Its Limitations

The implementation followed a classic technology-centric approach:

- Technical teams led design with clinical input limited to specific workflows

- Training focused on system transaction processing rather than patient care integration

- Standardized templates and protocols prioritized data capture over interaction quality

- Performance metrics emphasized documentation compliance and system utilization

While achieving technical implementation targets, this approach had created what clinicians called "digital disconnection", technology that captured information effectively but disrupted the human connections essential to healthcare. Patients experienced fragmented journeys despite unified technology, and care teams struggled to leverage the system for true coordination.

The Pivot to True Transformation

Three years into implementation, a new CMIO (Chief Medical Information Officer) and CNIO (Chief Nursing Information Officer) partnership initiated a fundamental redesign. Key shifts included:

1 Experience-First Redesign

Rather than centering design on technology capabilities, they mapped both patient and provider experience journeys to identify where technology was helping or hindering. This assessment revealed critical issues:

- Providers spent 43% of patient encounters focused on documentation
- Technology created artificial boundaries between care team members
- Patients received fragmented communication despite centralized information
- The cognitive flow of clinical work was frequently interrupted by system requirements

Based on this assessment, they redesigned core workflows around human experience rather than data creating "experience-optimized" templates, implementing team documentation approaches, and establishing technology-free zones in certain care contexts.

2 Team Capability Building

Recognizing that true transformation required collective capability rather than individual proficiency, they shifted from individual training to team capability development:

- Created care team simulation labs where entire units practiced technology integration

- Established peer coaching networks of clinician "digital guides"
- Implemented team-based optimization sessions focused on collaborative workflow
- Developed shared mental models for technology-enabled care coordination

This collective approach built the collaborative capabilities needed for technology to enable team-based care rather than individual documentation.

3 Adaptive Implementation

Rather than standardizing implementation across all settings, they adopted a context-sensitive approach that recognized different care environments had different needs:

- Created specialty-specific optimization teams led by practicing clinicians
- Implemented "90-day improvement cycles" where teams could evolve their approach
- Established variation governance that distinguished between harmful and beneficial differences
- Developed clinical decision support that adapted to patient and provider context

This adaptive approach allowed the same technology platform to serve diverse care settings effectively rather than imposing one-size-fits-all standardization.

4 Human Connection Design

Perhaps most importantly, they deliberately designed for technology to enhance rather than replace human connection:

- Created "relationship-preserving" documentation approaches (including collaborative documentation with patients)

- Implemented screen-sharing protocols that made technology a collaborative tool
- Designed room layouts that facilitated eye contact and presence during technology use
- Established technology-free moments in care journeys where human connection was prioritized

These human connection elements transformed technology from a barrier to an enabler of meaningful care relationships.

5 Intelligence for Improvement

Finally, they shifted analytics focus from compliance monitoring to continuous improvement:

- Created team-level dashboards showing how technology affected care outcomes
- Implemented patient feedback systems specifically addressing technology impact
- Established learning communities where teams shared technology optimization insights
- Developed predictive models that identified emerging workflow challenges before they created barriers

This improvement orientation transformed data from a judgment mechanism to a learning resource that enabled continuous evolution.

Outcomes and Insights

The reimagined approach delivered dramatically different results than the technology-centric implementation:

- Provider satisfaction with health information systems improved from 34% to 72%
- Patients reported 68% of the improvement in provider's presence during visits
- Care coordination metrics showed 47% improvement across transitions

- Documentation time decreased by 36% while quality metrics improved
- Clinical teams reported significantly reduced burnout related to technology use

Key insights from this transformation included:

1. Experience is the ultimate metric: Technical success means little if it degrades human experience for either providers or patients.

2. Team capability exceeds individual capability: Technology's impact depends more on collective patterns than individual proficiency.

3. Adaptive approaches outperform standardization: Allowing contextual adaptation within guiderails produces better outcomes than rigid standardization.

4. Human connection remains paramount: Technology creates most value when designed to enhance rather than replace essential human relationships.

5. Continuous evolution beats perfect implementation: Building ongoing improvement capability delivers more value than achieving initial perfection.

The CMIO reflected: "We initially treated this as a technical challenge of getting the right information to the right place at the right time. We eventually realized it was a human challenge of preserving and enhancing healing relationships in a technology-enabled environment. When we reframed our goal that way, everything changed for the better."

Case Study 4: Financial Services Customer Experience Transformation

Organization Profile: A national financial services provider offering banking, investment, and insurance products with 12,000 employees serving 3.5 million customers through both digital and physical channels.

Transformation Context: The organization had invested heavily in digital customer experience technologies but was seeing declining customer satisfaction, minimal adoption of new capabilities, and growing tension between digital and physical channels despite sophisticated technology implementation.

Technology-First Approach and Its Limitations

The organization had approached transformation from a primarily technological perspective:

- Digital and physical experiences were designed and managed separately
- Technology focused on transaction efficiency rather than relationship enhancement
- Implementation metrics emphasized feature deployment and utilization
- Training concentrated on system functionality rather than customer experience integration

This approach created what customers experienced as "digital fragmentation", sophisticated but disconnected capabilities that failed to create coherent journeys or meaningful relationships. Employees felt caught between pushing digital adoption and meeting customer needs, while leadership struggled to realize the strategic benefits they had expected from their technology investment.

The Pivot to True Transformation

After considerable internal debate, the organization fundamentally reset their approach. Key shifts included:

1 Journey Integration

Rather than treating digital and physical as separate channels, they mapped integrated customer journeys that spanned both realms:

- Identified moments where customers moved between digital and human interactions
- Created seamless handoffs that preserved context and relationship
- Designed consistent experiences regardless of channel
- Implemented shared metrics that encouraged cross-channel collaboration

This integration transformed fragmented interactions into coherent experiences that leveraged both digital efficiency and human connection.

2 Relationship-Centered Design

Instead of focusing primarily on transaction efficiency, they redesigned experiences around relationship development:

- Mapped the emotional journey alongside functional requirements
- Identified moments where human interaction created distinctive value
- Designed technology to enhance rather than replace relationship development

- Created experiences that built trust through both digital and human touchpoints

This relationship transformed how employees viewed and used technology—from transaction processing to relationship enablement.

3 Capability Ecosystem

The organization built a comprehensive capability ecosystem that developed both technical and relationship skills:

- Customer-facing employees developed "digital-physical integration" capabilities
- Digital team-built empathy and relationship understanding through customer immersion
- Leaders learned to balance efficiency and experience in performance expectations
- Cross-functional teams developed collaborative design capabilities

This balanced capability development enabled employees to leverage technology while maintaining the human elements customers valued most.

4 Cultural Evolution

Recognizing that their efficiency-focused culture created barriers to experience transformation, they deliberately evolved their cultural operating system:

- Shifted from efficiency-first to experience-first mindsets
- Redefined "digital success" from adoption metrics to relationship outcomes
- Created cross-functional accountability for integrated experiences
- Celebrated stories of technology-enabled human connection

These cultural shifts transformed how people thought about and applied technology throughout customer journeys.

5 Adaptive Learning Systems

Rather than defining the "perfect" experience design upfront, they created systems for continuous learning and adaptation:

- Implemented real-time customer feedback tied to specific journey moments
- Created rapid experimentation capabilities for testing experience innovations
- Established cross-functional "experience labs" where teams could prototype new approaches
- Developed systematic methods for scaling successful innovations across channels

This learning orientation transformed customer experience from a static design to a continuously evolving ecosystem.

Outcomes and Insights

The reimagined approach delivered significantly stronger outcomes than the technology-centric model:

- Customer satisfaction increased from 72% to 91% across integrated journeys
- Digital adoption increased 38% as customers gained confidence in cross-channel consistency
- Customer retention improved 14% while share-of-wallet metrics showed significant growth
- Employee engagement in customer-facing roles increased from 68% to 86%
- Cost-to-serve decreased 17% through more effective channel integration

Key insights from this transformation included:

1. Integration creates more value than optimization: Connecting experiences across channels delivered more impact than perfecting individual touchpoints.

2. Relationships remain fundamental: Technology created most value when designed to enhance rather than replace essential relationship elements.

3. Employee experience drives customer experience: How employees experienced and engaged with technology directly affected customer outcomes.

4. Adaptive approaches outperform perfect design: Building systems for continuous learning and adaptation delivered more sustainable value than pursuing perfect initial design.

5. Culture determines technology utilization: The same technologies created either transactional efficiency or relationship enhancement depending on the cultural context.

The Chief Customer Officer reflected: "We initially saw this as a digital transformation that would reshape customer experience. We eventually understood it was a human transformation enabled by digital capabilities. When we put relationships at the center rather than technology, everything else fell into place."

Cross-Case Insights: Patterns of True Transformation

Across these diverse case studies, several consistent patterns emerge that characterize true transformation beyond technology implementation:

1. Purpose Transcends Technology

In each case, transformation gained momentum when its purpose shifted from technological implementation to meaningful human and organizational outcomes. This purpose reframing connected change to intrinsic motivation rather than extrinsic compliance, creating engagement that technical mandates alone could never achieve.

2. Experience Design Precedes Technology Design

Organizations achieved breakthrough results when they designed desired human experiences first and then determined how technology could enable those experiences. This experience-first approach contrasts sharply with technology-first approaches that force human adaptation to technical constraints.

3. Work Redesign Creates Transformation Potential

In each case, meaningful transformation occurred when organizations fundamentally reimagined how work happened rather than simply digitizing existing processes. This redesign of work created possibilities that technology alone could not have delivered.

4. Collective Capability Exceeds Individual Proficiency

The most successful transformations invested heavily in building collective capabilities across teams and functions rather than focusing exclusively on individual technology proficiency. This collective approach enabled new patterns of collaboration and coordination that individual training could not have created.

5. Culture Determines Technology Value

In every case, the same technologies created dramatically different outcomes depending on the cultural context in which they operated. Organizations that evolved their cultural operating systems alongside their technological systems achieved results that technology-only transformations couldn't match.

6. Adaptive Approaches Outperform Planned Perfection

Organizations achieved superior results when they established mechanisms for continuous learning and adaptation rather than attempting perfect implementation of predetermined designs. This adaptive orientation transformed rigid initiatives into evolving ecosystems.

7. Integration Creates Exponential Value

The greatest transformation value emerged at integration points— where technologies connected across functions, where digital and human elements combined effectively, where data enhanced rather than replaced judgment. Organizations that deliberately designed for these integrations outperformed those that optimized isolated components.

From Case Studies to Your Transformation

As you reflect on these case studies, consider how their patterns might apply to your own transformation context:

1. **Purpose Assessment:** Does your transformation narrative emphasize technological implementation or meaningful human and organizational outcomes?

2. **Experience Orientation:** Are you starting with technological capabilities or desired human experiences?

3. **Work Reimagination:** Are you digitizing existing processes or fundamentally rethinking how work should happen?

4. **Capability Balance:** Are you investing as much in collective human capability as in technological capability?

5. **Cultural Alignment:** Have you assessed how your cultural operating system will enable or constrain technology's potential?

6. **Adaptive Capability:** Have you built mechanisms for learning and evolution throughout implementation?

7. **Integration Design:** Are you optimizing isolated components or designing for powerful integration points?

Your answers to these questions will significantly influence whether your technology investments deliver true transformation or merely expensive digitization. In the final chapter, we'll explore how to apply these insights to build your own transformation strategy beyond technology alone.

12

The Road Ahead: Building Your Transformation Strategy

CHAPTER 12

The Road Ahead: Building Your Transformation Strategy

Throughout this book, we've explored multiple dimensions that determine whether technology delivers transformative value or merely digitize existing patterns. We've examined how human psychology, organizational culture, leadership approaches, process design, capability building, change architecture, intelligence systems, HR's evolving role, and integrated transformation frameworks all contribute to true transformation beyond technology alone.

In this final chapter, we'll translate these insights into practical guidance for building your own transformation strategy. Whether you're just beginning a major initiative or seeking to redirect one already underway, these approaches will help you move beyond technology-centric transformation to create more holistic, sustainable change.

Assessing Your Current Approach

The journey toward true transformation begins with honest assessment of your current approach. The Transformation Orientation Diagnostic below provides a structured framework for this evaluation. For each dimension, consider where your organization currently falls to the continuum:

1. Purpose Orientation

- **Technology-Centric:** Focus on implementing and adopting new technological capabilities

- **Transformation-Centric:** Focus on achieving meaningful human and organizational outcomes enabled by technology

2. Design Starting Point

- **Technology-Centric:** Design begins with technological capabilities and requirements
- **Transformation-Centric:** Design begins with desired human experiences and work patterns

3. Scope Definition

- **Technology-Centric:** Primarily addresses technological systems and directly related processes
- **Transformation-Centric:** Addresses technological, human, cultural, and organizational dimensions as an integrated system

4. Leadership Approach

- **Technology-Centric:** Emphasizes directing and controlling implementation according to predetermined plans
- **Transformation-Centric:** Emphasizes creating conditions for continuous adaptation and emergent learning

5. Change Strategy

- **Technology-Centric:** Focuses on managing transitions from current to future states based on predetermined designs
- **Transformation-Centric:** Focuses on building adaptive capacity for continuous evolution based on emerging insights

6. Success Definition

- **Technology-Centric:** Primarily measures implementation milestones, adoption rates, and efficiency improvements
- **Transformation-Centric:** Measures holistic outcomes across human, operational, and strategic dimensions

This assessment provides a starting point for identifying where your current approach may be overly technology-centric and which dimensions require greater attention to enable true transformation.

The Strategic Reset: When and How

If your diagnostic reveals significant technology-centricity, you may need what I call a "strategic reset," a deliberate pivot from technology-focused to transformation-focused approaches. The case studies in the previous chapter illustrated how organizations successfully executed these pivots, often after initial disappointment with technology-centric results.

Key indicators that a strategic reset may be needed include:

1. Technology implementation is technically successful but delivering minimal performance improvement
2. Stakeholders are complying with but not engaging in the transformation
3. The organization is experiencing unexpected resistance or cultural challenges
4. The promised strategic benefits of technology remain elusive despite adoption
5. The human experience of technology is creating unintended negative consequences

If you recognize these patterns, consider these proven approaches to strategic reset:

1. Reframe Before Redesign

Before making structural changes, reframe the narrative around your transformation. Shift from technology-focused language ("implementing our new digital platform") to purpose-focused language ("creating seamless customer journeys" or "enabling more meaningful patient care").

This narrative shift isn't merely semantic; it fundamentally alters how stakeholders understand and engage with the initiative. A pharmaceutical company struggling with clinical trial technology adoption achieved a breakthrough simply by reframing the initiative from "implementing our new trial management system" to "accelerating life-changing treatments to patients."

2. Experience Before Efficiency

Temporarily set aside efficiency targets and refocus on the experience of those using and affected by the technology. Conduct structured assessment of:

- **Emotional journey:** How change affects psychological well-being and meaning
- **Cognitive load:** How technology affects mental bandwidth and decision quality
- **Relationship impact:** How technology influences connections between people
- **Purpose alignment:** How technology supports or hinders what people find meaningful

A financial services organization used this approach to reset their wealth management platform implementation. By mapping advisor and client experience journeys—and identifying where technology was creating friction in valued relationships—they redesigned their approach to preserve relationship elements while still gaining efficiency benefits.

3. Co-Create Before Mandate

Shift from top-down implementation to collaborative co-creation that engages those most affected by the change in designing how technology integrates with work. Effective approaches include:

- **Design partnerships** between technical teams and work practitioners
- **Prototype testing** with structured feedback from diverse users
- **Adaptation authority** for local teams to adjust implementation for their context
- **Innovation showcases** where teams share successful adaptations

A manufacturing organization used this co-creation approach to reset their production technology implementation. By establishing design teams where operators and engineers collaboratively redesigned work processes, they transformed technology from an imposition to an enablement platform that significantly improved both engagement and performance.

4. Capability Before Compliance

Shifts focus from adoption compliance to building the capabilities that enable effective technology utilization. This capability focus includes:

- Developing **adaptive skills** alongside technical proficiency
- Building **team capabilities** for collaborative technology use
- Creating **learning systems** that accelerate knowledge sharing
- Establishing **leadership capabilities** for technology-enabled environments

Healthcare organizations reset their electronic health record implementation by shifting from compliance monitoring to capability building. They created simulation labs where care teams practiced technological integration, established peer coaching networks, and developed team-based workflow optimization. This capability focuses dramatically improved both adoption quality and care outcomes.

Building Your Transformation Strategy

Whether you're planning a new initiative or resetting an existing one, these seven elements provide a framework for building a transformation strategy that goes beyond technology alone:

1. Develop Multidimensional Purpose

Create a transformation narrative that connects technological change to meaningful outcomes across multiple dimensions:

- **Human dimension:** How transformation will enhance experiences and capabilities
- **Operational dimension:** How transformation will improve work processes and outcomes
- **Strategic dimension:** How transformation will create distinctive organizational value
- **Social dimension:** How transformation will contribute to broader purpose

A retail organization embodied this multidimensional purpose in their customer experience transformation. Rather than focusing narrowly on digital adoption, they articulated how new capabilities would simultaneously enhance customer relationships, improve operational efficiency, create competitive differentiation, and advance their community impact goals.

2. Design Integrated Experiences

Rather than designing isolated technological touchpoints, create integrated experiences that span:

- **Human-technology interactions:** How people engage directly with systems
- **Human-human connections:** How technology affects relationships between people
- **Cross-channel journeys:** How experiences flow across digital and physical touchpoints
- **Emotional and functional dimensions:** How experiences feel as well as what they accomplish

A financial services firm used this integrated design approach for their client service transformation. They mapped holistic journeys spanning digital platforms, advisor interactions, and service team touchpoints—paying particular attention to transition points between channels and the emotional aspects of financial relationships.

3. Reimagine Work Fundamentally

Move beyond process digitization to fundamental work redesign that considers:

- **Purpose clarity:** What outcomes the work ultimately seeks to create
- **Value focus:** Which activities most directly contribute to those outcomes
- **Human-technology synergy:** How human and technological capabilities best complement each other
- **Connection patterns:** How work coordination and collaboration should evolve

A manufacturing organization applied this reimagination approach to their production operations. Rather than simply digitizing existing workflows, they fundamentally reconceived production work, shifting from rigid specialization to fluid team structures, from standardized procedures to guided adaptation, and from hierarchical control to distributed intelligence enabled by shared technology platforms.

4. Build Balanced Capability Ecosystems

Develop comprehensive capability building that integrates:

- **Technical proficiency:** Skills to operate new technological systems
- **Adaptive capacity:** Ability to learn, solve problems, and evolve approaches
- **Collaborative capability:** Skills for working together in technology-enabled environments
- **Leadership development:** Capabilities for guiding others through continuous change

A healthcare organization created this balanced ecosystem for their telehealth transformation. Beyond technical training, they developed clinical judgment in virtual contexts, built team capabilities for coordinating care across physical and digital touchpoints, and created leadership approaches that balanced standardization with adaptation as telehealth practices evolved.

5. Shape Cultural Enablers

Identify and deliberately develop cultural characteristics that enable technology to deliver transformative value:

- **Learning orientation:** Comfort with experimentation and evolution
- **Collaborative norms:** Patterns of working across boundaries
- **Trust dynamics:** Relationships between levels and functions
- **Decision practices:** How choices are made and by whom
- **Purpose connection:** How work links to meaningful outcomes

A financial services organization focused on these cultural elements during their analytics transformation. They created deliberate interventions to shift from information hoarding to sharing, from intuition-based to evidence-informed decisions, from risk avoidance to managed experimentation, and from functional silos to collaborative problem-solving.

6. Design Change Architecture

Create the conditions for successful adaptation rather than simply managing transitions:

- **Psychological safety:** Environments where people can express concerns and experiment

- **Learning mechanisms:** Structures for capturing and applying emerging insights

- **Participation platforms:** Opportunities for those affected to shape implementation

- **Resource flexibility:** Ability to adjust support based on emerging needs

- **Feedback systems:** Channels for identifying what's working and what isn't

A retail organization designed this architecture for their store technology transformation. They established innovative teams in each region, created digital community platforms for sharing emerging practices, implemented rapid feedback systems that identified adoption barriers, and maintained flexible implementation resources that could be deployed based on specific store needs.

7. Create Intelligence Systems

Develop mechanisms for transforming data into insight and action:

- **Integrated analytics:** Connecting data across traditional silos

- **Decision support:** Embedding intelligence into workflow moments

- **Learning focus:** Using data for improvement rather than judgment
- **Human-algorithm partnership:** Complementary use of machine and human intelligence
- **Ethical frameworks:** Guidelines for appropriate data use

A pharmaceutical company built these systems for their research transformation. They integrated data across previously separate research platforms, created decision support tools that enhanced scientific judgment, established learning forums where data patterns informed research strategy, and developed clear principles for algorithm use that maintained human scientific expertise at the core of discovery.

Orchestrating the Transformation Journey

Beyond these strategy elements, successful transformation requires sophisticated orchestration across multiple dimensions and time horizons. Key orchestration practices include:

1. Pace Layering

Rather than implementing all elements at the same speed, successful transformations utilize "pace layering" that recognizes different dimensions evolve at different rates:

- **Fast layers:** Technical configurations and process adjustments
- **Medium layers:** Skill development and work pattern evolution
- **Slow layers:** Cultural characteristics and identity elements

A healthcare organization applied this pace layering approach to their patient care transformation. They rapidly implemented core technical capabilities while allowing more time for care team adaptation, deliberately pacing cultural evolution over a longer horizon, and creating synchronization points where these layers could realign throughout the journey.

2. Alignment Rituals

Establish regular forums where stakeholders across dimensions can align their understanding and approaches:

- **Sensemaking sessions** where implementation teams share emerging lessons
- **Integration forums** where technical and organizational elements reconnect
- **Horizon scanning** that identifies upcoming interdependencies
- **Narrative alignment** that maintains consistent transformation story

A manufacturing organization created quarterly "transformation summits" where leaders from technology, operations, HR, and business units aligned their understanding of implementation progress, identified integration challenges, and coordinated upcoming activities. These rituals-maintained coherence across what could otherwise have become fragmented efforts.

3. Dynamic Resourcing

Move beyond fixed budgeting to more dynamic resource allocation that can respond to emerging needs:

- **Stage-gate funding** that releases resources based on learning rather than schedules
- **Flexible support pools** that can address unexpected challenges
- **Capability investment** that balances technical and human development

- **Value-based prioritization** that allocates resources based on emerging impact patterns

A financial services organization transformed their budgeting approach for digital transformation into this dynamic model. Rather than allocating all resources based on initial plans, they maintained 40% of their transformation budget in flexible pools that could be deployed based on quarterly impact assessment and emerging capability needs.

4. Ecosystem Leadership

Establish distributed leadership across the transformation ecosystem rather than centralized control:

- **Executive sponsors** who provide resources and remove barriers
- **Transformation guides** who design and orchestrate the overall journey
- **Local champions** who lead adaptation in specific contexts
- **Technical experts** who provide specialized knowledge
- **Integration leaders** who connect previously separate domains

A retail organization activated this leadership ecosystem for their customer experience transformation. They identified and developed leaders at multiple levels—from executive sponsors who shaped overall strategy to store-level champions who guided local implementation. This distributed approach enabled both strategic coherence and contextual adaptation.

Starting Where You Are

While the framework above provides a comprehensive transformation approach, most organizations face practical constraints that prevent the implementation of all elements simultaneously. The key is starting where you are with strategic interventions that create momentum for broader transformation.

Consider these high leverage starting points based on your current situation:

If You're Planning a New Initiative

- **Conduct a Transformation Readiness Assessment** that evaluates organizational readiness across all dimensions, not just technical requirements.

- **Establish a Multidimensional Governance Structure** that includes not only technical expertise but also experience in design, cultural, and leadership perspectives.

- **Create an Integrated Transformation Roadmap** that sequences technical, human, and organizational elements with appropriate interdependencies.

- **Develop a Balanced Measurement Framework** that includes leading indicators across human, operational, and strategic dimensions.

- **Build Transformation Capabilities** in your leadership team and implementation partners before technical deployment begins.

A pharmaceutical company used this approach when launching their research platform transformation. Before selecting technology vendors, they conducted comprehensive readiness assessments, established cross-functional governance, created an integrated roadmap spanning technology and culture, developed balanced metrics beyond technical milestones, and invested in building transformation capabilities in their leadership team.

If You're Mid-Implementation

- **Conduct an Experience Audit** to identify where technology is creating friction or enhancing human experience.

- **Establish Learning Forums** where implementation teams can share emerging insights and challenges across silos.

- **Create Integration Teams** focused specifically on connecting technical implementation with cultural and capability dimensions.

- **Implement Feedback Acceleration** systems that capture and respond to adoption challenges in real time.

- **Rebalancing Resources** between technical implementation and human/organizational dimensions as needed.

A healthcare organization used this approach to redirect their electronic health record implementation midway through deployment. They paused technical rollout for six weeks to conduct comprehensive experience assessments, establish cross-functional learning communities, create teams focused on workflow and culture, implement rapid feedback systems, and reallocate resources to previously under-addressed human dimensions.

If You're Post-Implementation But Pre-Transformation

- **Conduct a Value Gap Analysis** to identify where technology has been implemented but isn't delivering expected outcomes.

- **Launch Targeted Co-Creation** in areas where technology-work integration is problematic.

- **Develop Capability Accelerators** focused on building the skills needed to leverage existing technology more effectively.

- **Create Cultural Evolution Initiatives** addressing specific barriers to technology utilization.

- **Implement Value Realization Systems** that track and enhance outcomes beyond adoption of technology.

A financial services organization used this approach after their wealth management platform had been technically implemented but was delivering minimal value. They conducted structured gap analysis, established advisor co-creation teams in problematic areas, developed accelerated capability building for relationship-technology integration, launched specific cultural initiatives addressing data sharing barriers, and implemented comprehensive value tracking beyond system usage.

The Continuous Transformation Mindset

Perhaps most importantly, true transformation requires shifting from episodic change to continuous adaptation—recognizing that transformation isn't a discrete event but an ongoing capability in a world of accelerating technological and social evolution.

This continuous transformation mindset includes several key characteristics:

1. Learning Obsession

Rather than focusing primarily on implementation precision, continuous transformation leaders maintain relentless curiosity about what's working, what isn't, and why. They create structured learning mechanisms throughout their organizations and use these insights to continuously refine their approaches.

A healthcare organization embodied this learning obsession in their patient care transformation. They established weekly learning forums where teams shared emerging insights, created rapid feedback systems that identified adoption challenges, and maintained flexible implementation plans that evolved based on continuous learning.

2. Balanced Integration

Continuous transformation requires sophisticated integration across technological, human, and organizational dimensions. Leaders develop the ability to see interdependencies between these dimensions and create synchronization mechanisms that maintain alignment throughout continuous evolution.

A manufacturing organization developed this integration capability for their Industry 4.0 transformation. They created cross-functional teams responsible for technology-work integration, established regular forums where technical and organizational elements reconnected, and maintained integrated metrics that tracked how these dimensions evolved together.

3. Participatory Orientation

Rather than designing transformation for people, continuous transformation leaders design with people—creating structures for broad participation that tap into distributed intelligence while building ownership and engagement.

A retail organization demonstrated this orientation in their customer experience transformation. They established design partnerships between technical teams and customer-facing associates, created innovative showcases where teams shared successful adaptations, and maintained ongoing customer panels that provided experience feedback throughout implementation.

4. Systemic Awareness

Continuous transformation requires understanding organizations as complex adaptive systems rather than mechanical structures. Leaders develop the ability to identify leverage points, recognize emergent patterns, and work with rather than against systemic dynamics.

A financial services organization applied this systemic awareness to their digital banking transformation. Rather than forcing change through willpower and mandates, they identified cultural and structural enablers that would accelerate adoption, addressed systemic barriers like misaligned incentives and entrenched power dynamics, and created reinforcing feedback loops that made new behaviors self-sustaining.

5. Purpose Rootedness

Amidst continuous technological and organizational evolution, transformation leaders maintain strong connections to fundamental purposes the human and societal outcomes that technology ultimately serves.

A healthcare organization exemplified this purpose rootedness throughout their telehealth transformation. When facing difficult decisions about standardization versus customization, technical optimization versus human experience, or efficiency versus relationship, they consistently returned to their fundamental purpose of "creating healing connections regardless of distance" to guide their choices.

A Final Reflection:
The Human Core of Technological Transformation

As we conclude our exploration of transformation beyond technology alone, it's worth returning to the central insight that has run throughout this book: true transformation emerges from the integration of technological capability with human purpose, experience, and meaning.

The most sophisticated technologies—whether digital platforms, artificial intelligence, or whatever comes next—create value only to the extent that they enhance human capability, connection, and contribution. Organizations that recognize and design for this fundamental truth consistently outperform those that pursue technology for its own sake.

As you lead transformation in your own context, maintain this human-centered perspective. Ask not just "How can we implement this technology effectively?" but "How can we use this technology to create more meaningful work, stronger connections, better experiences, and greater value for those we serve?"

Technology alone is not transformation. But technology integrated with human purpose, capability, and connection can transform not just processes and outcomes but the very experience of work and its contribution to human flourishing.

The future belongs not to those with the most advanced technology but to those who most effectively integrate technological capability with human possibility. Your opportunity—and challenges to create transformation that achieves both.

13

Conclusion: The Continuous Transformation Mindset

CHAPTER 13

Conclusion:
The Continuous Transformation Mindset

As we conclude our journey through the dimensions of true transformation, it's worth reflecting on perhaps the most significant shift required of HR leaders and organizations today: embracing transformation not as a discrete project with a clear endpoint but as a continuous capability in a world of accelerating change.

The traditional transformation mindset views change as an exceptional state, a temporary disruption between periods of stability. Organizations planned major change initiatives, executed them (often painfully), and then sought to "freeze" new patterns before the next major disruption.

This episodic approach no longer serves us in an environment where technological, social, and market evolution is continuous rather than periodic. Today's reality demands what I call the "continuous transformation mindset", an orientation that views adaptation not as an occasional necessity but as a core organizational capability.

From Implementation to Evolution

The continuous transformation mindset shifts focus from perfect implementation of predetermined designs to guided evolution through continuous learning. This shift includes several key perspective changes:

From Change as Project to Change as Capability

Rather than organizing transformation as time-bound initiatives with clear beginnings and endings, organizations increasingly need persistent structures that enable continuous adaptation. These include:

- Ongoing innovation forums where emerging possibilities are explored
- Permanent cross-functional teams focused on experience evolution
- Embedded transformation roles within operational structures
- Dedicated resources for continuous adaptation rather than one-time change

A financial services organization exemplified this shift by establishing what they called "transformation platforms" rather than programs. These platforms—including innovation labs, capability accelerators, and integration forums—provided enduring infrastructure for continuous adaptation rather than time-bound implementation structures.

From Perfect Design to Continuous Prototyping

Rather than attempting to design perfect solutions before implementation, continuous transformation embraces ongoing experimentation through:

- Rapid prototyping of potential approaches
- Small-scale pilots that generate learning before broader deployment
- Progressive implementation that evolves based on emerging insights
- Comfort with "perpetual beta" as a philosophical stance

A healthcare organization applied this approach to their telehealth evolution. Instead of designing a comprehensive telehealth model upfront, they created a continuous prototyping system where clinical teams tested different approaches, captured structured learning, and progressively evolved their virtual care model based on both clinical outcomes and human experience.

From Resistance Management to Co-Evolution

Rather than viewing stakeholders as sources of resistance to be managed, continuous transformation sees them as co-creators in an evolving system:

- Engaging those affected by change in designing new approaches
- Creating feedback systems that capture diverse perspectives
- Establishing distributed innovation that harnesses collective intelligence
- Building shared ownership through participatory processes

A manufacturing organization transformed their approach to production technology implementation from centralized design to co-evolution. They established operator-engineer partnerships that continuously refined technology integration, created cross-plant communities that shared emerging practices, and maintained ongoing dialogue between technology providers and production teams that influenced both implementation and future development.

Leading Continuous Transformation

This shift from episodic to continuous transformation requires new leadership approaches that balance direction with emergence, stability with adaptation, and efficiency with learning.

Key leadership practices include:

1. Evolutionary Purpose

Rather than defining rigid end states, continuous transformation leaders articulate what I call "evolutionary purpose"—directional aspiration that provides guidance while allowing for multiple pathways and emergent possibilities.

A retail organization exemplified this approach in their customer experience transformation. Rather than specifying exact future state designs, leadership articulated an evolutionary purpose of "creating seamless, personalized experiences that build lasting relationships" and established guiding principles that provided direction while enabling continuous innovation in how that purpose was fulfilled.

2. Polarity Management

Continuous transformation inevitably involves tensions between competing values: standardization versus customization, efficiency versus experience, stability versus innovation. Effective leaders develop skill in what I call "polarity management", the ability to navigate these tensions as ongoing polarities to be balanced rather than problems to be solved.

A healthcare organization demonstrated this skill in managing the tension between standardized protocols and clinical judgment in their care transformation. Rather than choosing one over the other, they created sophisticated approaches that provided standardized foundations while preserving space for professional judgment and continuous evolution of the standards themselves based on clinical insight.

3. Network Activation

Rather than relying primarily on hierarchical authority, continuous transformation leaders activate networks that enable change to flow through social connections. This activation includes:

- Identifying and engaging informal influences
- Creating connection points between previously separate networks
- Establishing communities of practice around transformation challenges
- Developing backbone structures that support network effectiveness

A global manufacturer applied this network approach to their digital transformation. They mapped influence networks across plants, created "transformation communities" that connected digital champions across locations, established virtual forums where these networks could share emerging practices, and provided support infrastructure that enabled these networks to drive change more effectively than top-down approaches alone could have achieved.

4. Learning Acceleration

Perhaps most critically, continuous transformation leaders create what I call "learning acceleration"—the systematic capability to capture, distill, and apply insights faster than the pace of change itself. This acceleration includes:

- Designing for rapid feedback from multiple perspectives
- Creating reflection practices that extract meaning from experience
- Establishing knowledge flows that spread insights across boundaries
- Building experimental structures that generate validated learning

A professional services firm exemplified this learning acceleration in their digital workplace transformation. They established rapid feedback systems that captured emerging usage patterns, created weekly reflection forums where teams extracted insights from these patterns, developed knowledge-sharing platforms that rapidly diffused effective practices across the organization, and maintained an experimental portfolio that continuously tested new approaches to virtual collaboration.

HR's Evolving Role in Continuous Transformation

For HR leaders, the shift to continuous transformation requires evolving beyond both traditional change management and even the transformation agent roles described in earlier chapters. It requires becoming what I call "adaptation architects", professionals who design and maintain the conditions for continuous organizational evolution.

This evolving role encompasses several key dimensions:

1. Capability Ecosystem Stewardship

Rather than focusing primarily on discrete learning programs, HR creates and maintains comprehensive capability ecosystems that enable continuous adaptation:

- Integrated development experiences spanning formal, social, and experiential learning
- Knowledge networks that connect expertise across boundaries
- Talent mobility that spreads capability through experience
- Leadership development focused on adaptive capacity

A technology company's HR function exemplified this stewardship in their approach to building digital collaboration capability. They created an integrated ecosystem comprising curated learning resources, peer coaching networks, project-based skill development, cross-functional rotation experiences, and leadership development focused specifically on guiding distributed teams through continuous technology evolution.

2. Architecture for Emergence

HR increasingly designs organizational structures, processes, and systems that enable emergence rather than control—creating conditions where innovation and adaptation can flourish:

- Flexible organizational designs that can reconfigure as needs evolve

- Decision architectures that distribute authority appropriately

- Collaboration infrastructures that enable boundary-spanning work

- Feedback systems that accelerate adaptation and learning

A professional services firm's HR team transformed their approach from designing static organizational structures to creating what they called "adaptive architectures", frameworks that provided sufficient structure for coordination while enabling teams to reconfigure around emerging client needs and technology possibilities.

3. Cultural Evolution Guidance

Rather than one-time culture change initiatives, HR provides ongoing guidance for cultural evolution that enables continuous adaptation:

- Identifying cultural characteristics that enhance or inhibit adaptability

- Designing interventions that shape cultural evolution in desired directions

- Creating feedback systems that track cultural patterns and their impacts

- Advising leaders on cultural implications of technological and strategic choices

A healthcare organization's HR function established what they called a "culture observatory" that monitored how their culture was evolving during their digital transformation. This observatory combined quantitative measures like collaboration patterns and psychological safety indicators with qualitative assessment through story capture and dialogue analysis, providing rich insight into how cultural factors were enabling or constraining adaptation.

4. Human Experience Design

As technology increasingly enables continuous reconfiguration of work, HR leads in designing human experiences that maintain meaning and connection amidst ongoing change:

- Mapping how technological evolution affects psychological and social experiences
- Designing transitions that preserve meaningful aspects of work amidst change
- Creating rituals and practices that maintain human connection
- Ensuring technology serves human flourishing rather than the reverse

A financial services organization's HR team led a deliberate experience design initiative alongside their agile transformation. They mapped how changing work patterns affected sense of identity and belonging, designed new connection rituals that maintained relationships amid changing team structures, and established experience principles that guided technology implementation decisions.

5. Ethical Evolution Navigation

Perhaps most critically, as transformation raises complex ethical questions about how technology affects human work and well-being, HR increasingly provides ethical guidance for navigating these territories:

- Articulating principles for human-centered technology integration

- Facilitating dialogue about ethical implications of automation and augmentation
- Designing approaches that balance efficiency with meaningful human contribution
- Advocating technology that enhances rather than diminishes human potential

A manufacturing organization's HR leadership established an "Ethical Work Evolution Forum" that brought together leaders from operations, technology, and HR to address complex questions emerging from their advanced automation initiative. This forum developed principles for human-machine collaboration, created assessment processes for automation decisions, and established ongoing dialogue about how technology could enhance rather than replace meaningful human work.

Building Your Continuous Transformation Capability

As you consider how to develop continuous transformation capability in your organization, consider these starting points:

1. Assess Your Transformation Metabolism

Just as individuals have different metabolic rates, organizations have different "transformation metabolisms", inherent capacities for absorbing and integrating change. Assess yours honestly considering:

- Recent change history and residual fatigue
- Cultural openness to novelty and experimentation
- Leadership comfort with ambiguity and emergence
- Structural flexibility and boundary permeability
- Learning capability and knowledge flows

This realistic assessment establishes a baseline for developing transformation capability without overwhelming existing capacity.

2. Create Transformation Infrastructure

Establish enduring structures that enable ongoing adaptation rather than relying solely on temporary project teams:

- Innovation forums where emerging possibilities are explored
- Integration of mechanisms that connect technological and human dimensions
- Learning systems that capture and apply insights across boundaries
- Capability accelerators that build adaptive skills throughout the organization

A pharmaceutical company created this infrastructure for their research transformation through a network of innovation labs embedded within research teams, quarterly integration forums that connected technology and scientific perspectives, a digital knowledge common that captured emerging insights, and a capability academy focused specifically on building digital research skills.

3. Develop Second-Order Capabilities

Beyond specific skills for current technologies, build what I call "second-order capabilities", meta-skills that enable continuous adaptation to whatever comes next:

- Learning agility: Ability to rapidly acquire new knowledge and skills
- Systems thinking: Capacity to understand complex interdependencies
- Design thinking: Skill in human-centered problem solving
- Network leadership: Ability to influence through connection rather than control
- Polarity navigation: Capability to balance competing values effectively

A technology company built these capabilities through a comprehensive leadership development initiative focused not on managing current change but on building the adaptive capacity for continuous evolution. This development included immersive learning experiences in complex problem solving, cross-boundary collaboration projects, and ongoing coaching focused specifically on navigating the tensions inherent in continuous transformation.

4. Establish Reflection Disciplines

Perhaps most fundamentally, create disciplines for regular reflection that extract meaning and direction from the continuous flow of experience:

- Team reflection rituals that capture learning from implementation
- Cross-functional forums that identify patterns across separate initiatives
- Leadership practices that deliberately pause for sensemaking
- Organizational processes that translate reflection into adaptation

A professional services firm established these disciplines through what they called "transformation learning cycles", structured practices for extracting insights at multiple levels, from weekly team reflections to quarterly organizational pattern analysis. These disciplines dramatically accelerated their ability to learn and adapt throughout their digital transformation journey.

Transformation Without End

As we close our exploration of transformation beyond technology alone, it's worth acknowledging a fundamental truth: the journey we're discussing has no destination. We live in an age of continuous technological, social, and organizational evolution that requires ongoing adaptation rather than one-time change.

This reality can seem exhausting when viewed through the lens of traditional transformation approaches that treat change as exceptional and stability as normal. But when we develop the mindsets, capabilities, and infrastructures for continuous transformation, adaptation becomes not an exhausting disruption but an energizing opportunity to create ever more effective, humane, and meaningful organizations.

The technologies that enable this continuous transformation, from artificial intelligence to collaborative platforms to whatever comes next—are merely tools. Their transformative potential depends entirely on how skillfully we integrate them with human purpose, capability, and connection.

As you lead transformation in your own context, remember this essential truth: Technology alone is not transformation. True transformation emerges when we harness technological possibilities to enhance human potential, creating organizations where technology serves people, purpose, and possibilities previously unimagined.

The future belongs to those who can navigate this continuous evolution not as a burden to be managed but as an opportunity to create organizations that are simultaneously more capable and more human. I hope this book has provided frameworks, insights, and inspiration for your journey toward that future.

www.ingramcontent.com/pod-product-compliance
Lightning Source LLC
Chambersburg PA
CBHW040852210326
41597CB00029B/4815